GLOSSARY

OF

MICROSCOPICAL

TERMS

AND

DEFINITIONS

SECOND EDITION, 1989

NEW YORK MICROSCOPICAL SOCIETY
15 West 77th Street
New York, New York 10024

Editorial Committee

Walter W. Aschoff, Editor of First Edition

Lawrence Kobilinsky, Ph.D.

Roger P. Loveland, D.Sc.

Walter C. McCrone, Ph.D.

Theodore G. Rochow, Ph.D., Chairman

Acknowledgements

On behalf of the New York Microscopical Society, the Editorial Committee expresses profound gratitude to all those who helped with this glossary.

FOREWORD

"Words, like eyeglasses, blur everything that
they do not make clear."

Joseph Joubert (1754-1824)

The purpose of this glossary is to clarify the meanings of microscopically oriented terms that are not adequately defined in our dictionaries. One such example is the word pair *microscopic* and *microscopical*, which are equivalent in most, if not all, dictionaries. Yet they are quite opposite in meaning. *Microscopic* is an adjective describing an object too small to be resolved by the eye. *Microscopical* is the adjective which describes anything related to the microscope.

All kinds of microscopes have optics, so the term *optical microscope* is confusing and redundant. On the other hand, the term *light microscope* is very specific. *Photomicrograph* is another clear term, referring to the recording of microscopic details photographically. A *microphotograph* is a very small photograph (perhaps of a large article). Science writers in all disciplines should be careful to make such important distinctions, and with the help of this glossary, that task will be made easier.

PREFACE

T*he New York Microscopical Society*, incorporated in 1877, is one of the oldest in America. It is an independent, non-profit organization supported by the dues of its members and by gifts and bequests.

THE PURPOSE OF THE SOCIETY IS THE ADVANCEMENT OF APPLIED AND THEORETICAL MICROSCOPY. In pursuit of this goal, it functions:

- To provide a meeting place and center for the dissemination of information on all phases of microscopy. Regular meetings are held in Room 419 of the American Museum of Natural History, Central Park West at 79th Street, New York City, at 8:00 PM on Fridays. Meeting notices are mailed to all members indicating the subject of the program and changes, if any, as to dates and places of meetings and any additions thereto.

- To promote knowledge of the proper use of the microscope and accessory equipment.

- To encourage the practice of microscopy in all fields of known usefulness as well as in areas not yet explored, so that its full potential may be realized in scientific study and practical application.

- To keep informed about, and encourage development of, new instruments and techniques.

- To preserve the historical instruments, literature, and work which have accumulated at this center.

- Not the least of its functions is to bring together, for mutual enjoyment and stimulus, professionals and amateurs with a kindred interest in *MICROSCOPY*.

Following a brief business session, a meeting typically comprises a lecture by a prominent microscopist, a panel discussion with audience participation, or a special cinema, video or slide show. The programs cover many fields of application, representing the varied interests of the members: biology, chemistry, ceramics, crystallography, forensic science, medicine, metallurgy, mineralogy, paper, textiles, wood, optics, and instrumentation. The emphasis of the programs is on light microscopy, with other advanced special instrumentation as additional areas of concern. The members are notified about synoptic courses in microscopy as the information becomes available. *WORK-SHOPS* are included in the programs of the Society to acquaint members with special techniques in microscopy and their practical application. Microscopes, lamps, and accessory apparatus are available for use by members and for workshops.

Other activities of the Society include the biennial *SYMPOSIA*, providing the opportunity for microscopists to keep abreast of changes in the science and to view and understand the latest and most sophisticated developments in instrumentation. Also included in the Society programs are field trips which include visits to laboratories and to outdoor areas for specimen collection.

The Society has an extensive library of historical, rare and contemporary books on subjects of interest to microscopists. With a few exceptions, books may be borrowed by members. Rare and old, as well as usable microscopes and slide collections, are available in our meeting room in the American Museum of Natural History.

A newsletter, NYMS NEWS, is published monthly. It contains announcements of meetings, information of general interest to members, abstracts of past meetings, listings of equipment wanted and available, book reviews, and miscellaneous items about our members. MICROSCOPIUM, issued semiannually, contains articles by members of the Society.

If you are interested in any phase of MICROSCOPY, whether it be research, routine analysis, the formation and interpretation of images, or simply the examination of objects for curiosity's sake, you will find the work of the Society and association with its members to be worthwhile.

We invite your membership and look forward to greeting you at our meetings. For your convenience an application blank is appended at the end of this volume.

GLOSSARY OF MICROSCOPICAL TERMS AND DEFINITIONS

Abbe, Ernst (1840-1905). German mathematician and physicist, professor at Jena, and inventor of much optical apparatus at the Zeiss works. His inventions include the apochromatic objective, the compensating ocular, the Abbe condenser, a well-corrected oil-immersion achromatic condenser, the immersion objective, Abbe apertometer, Abbe refractometer, and the drawing camera; he evolved the Abbe theory of resolution and microscope imagery, the numerical-aperture formula, and other optical theories.

Abbe apertometer. In microscopy, a device for measuring simultaneously, the numerical and angular apertures of an objective or condenser. The back focal plane of the objective is viewed with an auxiliary lens, and the device is set to show the position of an indicator just cutting into two opposite edges of the conoscopic field of view.

Abbe condenser. A two-lens system, uncorrected and having only a very low-angle aplanatic cone.

Abbe's law of limiting resolution. For a periodic structure of units separated by distance d and obliquely illuminated by the unrefracted ray and one of the two diffracted rays (extremely oblique illumination), Abbe applied the law of diffraction: $d = 0.5\ \lambda/NA$ where:

λ = wavelength of the monochromatic light or shortest of mixed wavelengths

NA = the limiting numerical aperture (*q.v.*) of objective or condenser.

See "Abbe theory of image formation."

Abbe substage apparatus. Includes a rack and pinion for horizontal displacement of an iris diaphram to obtain oblique lighting.

Abbe test plate. A long, wedge-shaped coverslip about 0.20 mm thick at one end and 0.10 to 0.12 mm at the other end coated chemically with a silver film on which are ruled horizontal lines. At each variation in thickness of 0.01 mm there are vertical lines. By means of oblique illumination and by focusing on different portions of the plate, it is possible to determine the optimum coverslip thickness for any objective and also, for microscopes with drawtubes, the tubelength for best objective performance. The approximate freedom from spherical and chromatic aberrations can also be estimated. Small isolated bits of silver near the edges of the lines form good objects for the star test (*q.v.*).

Abbe theory of image formation. Abbe's theory is based on the fact that a non-self-luminous particle, which is illuminated by an extraneous source, gives rise to diffracted light rays, in addition to the dioptric pencil. He stated that to form a good microscopical image as many of the diffracted rays as possible should be intercepted by the objective. With closely ruled lines, his theory is easily demonstrated by observing the back lens of the objective, for here the diffracted rays can be observed directly if the aperture diaphram is closed. It can be shown that, when the illumination is arranged to exclude the diffracted images, resolution is lost.

aberration. In an optical system, any defect that degrades its performance from that which could be achieved with a perfect lens, *e.g.*, failure of an object point to be imaged as a point.

aberration, chromatic. A defect in a lens or optical system due to the greater refraction of shorter wavelengths over that of longer ones at a lens surface. Hence the focal length of a simple lens is shorter for blue than for red rays. This dispersion of the wavelengths will cause color fringes in the image field of a lens with such an aberration.

aberration, spherical. A lens defect whereby image-forming rays of one color, passing through the outer zones of a lens come to focus at a different distance from the lens than do those of more central rays. With a simple spherical (or plano-spherical) lens the outer rays always meet the axis closer to the lens than do more central rays and the lens is uncorrected or "undercorrected". When the reverse is true the lens has been "overcorrected".

achromatic. Literally, color free. A truly achromatic lens would transmit light without decomposing it spectrally and there would be no color fringing in the image. A doublet, composed of a positive and negative lens element, can be made achromatic for two colors which much improves the performance for most of the others.

acicular. Having a needle-like shape.

acute bisectrix (Bx$_a$). The bisector of the acute optic axial angle of biaxial crystals (_q.v._).

air-lock. An intermediate, enclosed chamber of a vacuum or pressure system through which an object may be passed without effectively changing the vacuum or pressure of the main body of the system. Reference here is to the vacuum of all electron microscopes.

Airy, Sir George Biddell (1801-1892). English astronomer and mathematician. Originator of the Airy theory of resolution. See Airy disk.

Airy disk. The image of a bright point-object, as focused by a lens system. With monochromatic light, it consists of a central point of maximum intensity surrounded by alternate circles of light and darkness caused by the reinforcement and interference of diffracted rays. The light areas are called maxima and the dark areas minima. The distribution of light from the center to the outer areas of the figure was investigated mathematically by Airy. The diffraction disk forms a basis for determining the resolving power of a lens system. The diameter of the disk depends largely on the aperture of the lens. The diffraction of light causing the Airy disk is a factor limiting the resolution in a light-optical system.

alpha (α). Greek letter representing the lowest of the three principal refractive indices of a biaxial crystal (*q.v.*); also the angle between the *b* and *c* axes of a triclinic crystal.

alpha-monobromonaphthalene. An organic liquid with a high refractive index, about 1.66. It is used in liquid mixtures for determining the refractive indices of solids. It is also used with the 1.6 NA metallographic objective. However, this objective may no longer be available.

Amici, Giovanni Battista (1784-1863). An Italian astronomer who made several contributions to optical science, including the Amici prism (*q.v.*).

Amici prism. A prism, invented by Amici, which disperses light without appreciable deviation and so can be used to make a direct-vision spectroscope. Amici prisms are used in certain spectroscopic eyepieces and in some Abbe refractometers.

amorphous. Lacking a repeating array of crystal lattice points; amorphous materials are isotropic but may show strain birefringence.

amplifier (optical). When the projected image of the microscope is projected less than 250 mm the objective may be moved enough from its usual position (for visual observation) to degrade the image. It can be restored by adding a lens above the ocular to bring the image into focus on the desired plane but with the objective still in its position of best-corrected image. A similar system requires an "ocular" in which the same corrections are made. Such lens systems are termed "amplifiers" (*e.g.*, Homals of Zeiss, Amphiplans of B & L, etc.).

analyzer. A second polarizing element inserted beyond a preparation. When its vibration direction is at right angles to the vibration direction of the polarizer, the field becomes black if no anisotropic specimen is on the stage or when viewing an anisotropic substance in an extinction position or directly down an optic axis of an anisotropic crystal. (See polarizer)

angstrom (Å). Unit of linear measure named after A.J. Ångström.

It is 1×10^{-10} meters; 1 micrometer = 10,000 Å. It is variously abbreviated A, A., A.U., Å., or ÅU. According to the Systeme Internationale (*q.v.*), the angstrom unit is obsolete; the equivalent is 0.1 nanometer (*q.v.*).

angular aperture (AA). The angle subtended between the axis of a lens and the largest accepted angle of the image-forming rays. With microscope objectives the trigonometric sine of this angle is used to define numerical aperture NA (*q.v.*) but as measured from the axial object point. (Note: Some commercial advertising literature has erroneously used the full angle in defining NA.)

anhedral crystals. Those whose growth has been impeded by adjacent crystals growing simultaneously, so that the development of plane faces is inhibited. Also crystals eroded, partly dissolved or mechanically deformed to the point where nearly all traces of crystal faces have been removed. All anhedral crystals are irregularly shaped and do not have plane faces, *cf.*, euhedral.

anisotropic. Quality of a transparent material having different refractive indices depending on the vibration direction of the transmitted light, hence any material that affects polarized light differently according to its direction through the material.

annular illumination. The result of placing a stop in the first focal plane of the condenser to produce an illuminating cone of light with all the light flux near the surface of the cone. The central part of the cone will be dark. This arrangement is often used for a darkfield effect in low-power work by transmitted light. The condenser must be correctly focused and centered. By reflected light, darkfield illumination is attained with an annular condenser fitted around the objective for opaque objects. Annular illumination implies that the object is lighted from all sides.

annular stop. The opaque ring-shaped stop with a small central opening usually placed in the objective back focal plane to provide dispersion staining.

aperture. In optics an aperture is an opening that restricts the size of the light beam that enters or leaves a lens or lens system and is often controlled by an iris aperture diaphram. Its correct placement can be a very critical matter. The aperture can, of course, be filled by a lens; it can be at a focal plane of the lens as is that of the substage condenser of a microscope. Opening this aperture wider increases image brightness and definition (resolving power) but decreases depth of field and contrast. Aperture size is measured variously according to the field of use of the lens, often by its simple diameter as is that of the telescope. (But see: angular aperture, relative aperture, numerical aperture.) The free aperture of a lens is that which is unrestricted except by its own diameter.

aperture (for electron microscopy).

anode aperture. The opening in the accelerating voltage anode shield of the electron gun through which the electrons must pass to irradiate the specimen.

condenser aperture. An opening in the condenser lens controlling the number of electrons entering the lens and the angular aperture of the electron beam. The angular aperture can also be controlled by the condenser lens current.

physical objective aperture. A metallic diaphram, with a small central hole, used to limit the cone of electrons accepted by the objective lens. This improves image-contrast since highly scattered electrons are prevented from arriving at the Gaussian image plane and therefore cannot contribute to background fog.

aplanatic. Free from spherical aberration and coma.

aplanatic points. The two conjugate points for which a completely spherical glass lens is aplanatic. They lie on the principal axis, both within the sphere.

apochromatic objective. An objective corrected for spherical aberration at two wavelengths and for chromatic aberration at three wavelengths.

artifact. A spurious image which does not correspond to the true microstructure of the object. For example, a source of artifacts may be the improper preparation of a substance for examination. The term artifact is sometimes used to refer to any object detail not germane to the problem at hand (*e.g.*, air bubbles or extraneous dust).

aspheric lens. A lens made aplanatic by grinding the outer zones to a greater radius than the inner zones. Aplanatic condensers can be made in this way. A simple aspheric lens is widely used as a lamp condenser because of its efficiency in converging light rays to one focal plane.

astigmatism. A defect in a lens or optical system which causes rays in one plane parallel to the optical axis to focus at a distance different from those in the plane at right angles to it. Astigmatism is equivalent to placing a weak cylindrical lens against a normal lens.

astigmatism of the eye. A condition in which the refracting surfaces of the eye are not truly spherical. Dots appear as lines, and lines radiating from a common center appear sharply only in certain azimuths. Astigmatism of the eye requires wearing corrective eyeglasses even while using a microscope if the best possible image is to be obtained. In turn, the ocular(s) should possess high *eyepoint(s)* (*q.v.*) for eyeglass wearers. Simple *myopia* (*q.v.*) or *hyperopia* (*q.v.*) does not require eyeglasses or eyepieces with high eyepoints.

ASTM. American Society for Testing and Materials, 1916 Race St., Philadelphia, PA 19103. For ASTM national standard definitions and nomenclature see References 2-5.

axis, crystallographic. One of several imaginary lines assumed in describing the positions of the planes by which a crystal is bounded, the positions of the atoms in the structure of the crystal and the directions associated with vectorial and tensorial physical properties.

axis, optical. Usually refers to the axis on which several principal lens axes may lie. It also refers to the axis of the eye which extends through the center of the eye lens. Not to be confused with optic axis (*q.v.*).

axis, principal. A line conceived as passing through the center of a lens to connect the centers of curvature of the lens surfaces. The focal points of a lens lie on the principal axis.

back focal length. As measured on the principal axis, from the second lens' vertex to the back focal point of the lens. It is not the equivalent of the focal length.

back focal plane. The plane, normal to the lens axis, situated at the back focus of a lens.

back lens. In any compound lens (a lens system composed of more than one lens element), the last lens through which the light passes is called the back lens. It may be a single simple lens, a doublet, or triplet. See front lens.

balsam, Canada. A resin from the balsam fir Abies balsamea. Dissolved in xylene, toluene, or benzene it is used as a mountant for permanent microscopical preparations. Its refractive index may vary from 1.530 to 1.545 and its softening point from room temperature to 100°C, these properties varying with age and solvent content. If impure it discolors with age. As solvent evaporates, the mountant shrinks, often leaving light-scattering voids.

base, microscopical. The microscope's supporting structure such as the horseshoe type. Heavier bases, some containing the entire illuminating system, began to appear after World War II.

Becke line. When the liquid phase of a microscopical mount has a refractive index different from that of the solid phase, a line or narrow band of light can be observed around or just within the outlines of the specimen as the microscope tube is raised or lowered from its position of best focus. The presence of the line indicates the difference in index referred to, and its absence, therefore, indicates similarity of index between the specimen and its mounting fluid. The Becke line is useful in determining the refractive index of transparent, microscopic particles. See refractive index by Becke line.

bellows length. The distance from the eyepoint to the image plane in a photomicrographic apparatus.

benzene (also benzol) C_6H_6. A liquid with a boiling point of 80.1°C, very slightly soluble in water. It should not be confused with benzine (*q.v.*).

benzine (also petroleum ether). A mixture of the lighter fractions of aliphatic hydrocarbons from refining petroleum. Its boiling range is lower than that of benzene, in some grades between 40° and 60°C. Because of its fast evaporation it is excellent for cleaning objective lenses.

Bertrand lens. A small, low-power lens, usually on a slide for insertion into the drawtube between analyzer and ocular. It is used to observe the back focal plane of the objective so as to examine interference figures or as an aid in achieving interference figures. It is apt to be strongly astigmatic. It is used to image the lamp filament in setting up Köhler illumination (*q.v.*) as well as for centering dispersion staining stops to the substage aperture diaphram.

beta (β). Greek letter representing the intermediate principal refractive index in a biaxial crystal (*q.v.*), also used to designate the angle between the *a* and *c* axes of a monoclinic or triclinic crystal.

biaxial crystals. Anisotropic crystals in the orthorhombic, monoclinic and triclinic systems. They have three principal refractive indices α, β and γ and two isotropic directions, *i.e.*, optic axes.

bicentric condenser. A darkfield condenser with two reflecting surfaces, sometimes called a bispheric condenser.

bifilar eyepiece. An ocular with two crossed hairs, wires, filaments or threads, each of which has perpendicular motion.

binocular microscope. A microscope fitted with double eyepieces for vision with both eyes. The purpose in dividing the <u>same</u> image from a single objective of the usual compound microscope is to reduce eyestrain and muscular fatigue which may result from monocular, high-power microscopy. The purpose in obtaining a <u>different</u> image for each of two oculars is to provide stereoscopy by means of two different angles of view. There are two kinds of stereoscopic microscopes: binobjective (Greenough) older type and monobjective (common main objective, CMO) newer type. (See stereomicroscope, Greenough, etc.)

birefringence. The numerical difference in refractive indices for a substance. In a given crystal view, the interference color (retardation) between crossed polars depends on the birefringence and thickness:

Retardation (nm) = 1000 x thickness (μm) x birefringence

bispheric condenser. See bicentric condenser.

black-body radiation. See color temperature.

body-color. The color of a transparent body as seen by transmitted light. It is not necessarily the same as surface color (*q.v.*).

brightfield illumination. The method of lighting the specimen with a solid cone of rays. Transmitted brightfield illumination is performed by a substage condenser. Reflected brightfield illumination is performed by a vertical illuminator (*q.v.*). See also condenser, brightfield. Compare darkfield illumination.

brightness. The brightness of an extended luminous source is termed the intensity per unit area measured in candles per cm^2. Also, the intensity of reflection; it may be measured in lamberts; one lambert equals 1 lumen/cm^2.

brownian motion. The random oscillatory movements of microscopic particles suspended in a mobile liquid which are caused by the thermal oscillations of the molecules of the liquid. This effect, discovered by the English botanist Robert Brown, is easily observed microscopically with suspended particles.

bull's eye. Usually a simple lens, either double convex or planoconvex. The plane side should be toward the light pencil of greatest angle. The lens is of low power; as used in microscopy it generally has a focal length of about 4 inches. Used as a light-collecting lens at the lamp or for spotlighting objects from above the stage. Well-corrected doublets of long focus may be available.

camera lucida. A partially reflecting prism-mirror device fitted over the microscope ocular, enabling the eye to see simultaneously the field of view and any object alongside the microscope.

camera, photomicrogaphic. A container to hold photographic film or plates with the film protected by a film holder and shutter, for photographing microscopical images.

cardioid darkfield condenser. A condenser designed with two reflecting surfaces; the first, a spherical surface which reflects the rays to a second, cardioid (heart-shaped) surface. The virtue in such an arrangement is that, if the cardioid surface is of true* figure, the lens is both achromatic and aplanatic. It has a limiting numerical aperture of about 1.0. Thus objectives of a greater numerical aperture cannot be used successfully with it.
*A "true" cardioid figure is the trace of a point on the circumference of a circle rolling around an equal, fixed circle.

cassegrainian darkfield condenser. Named after Cassegrain, astronomer of the 17th century. A high-power, darkfield condenser to be used with objective apertures as high as 1.3. It is sometimes called the luminous spot ring condenser (Zeiss).

cassette. A light-tight film or plate holder.

CAT. Computer-aided tomography uses computers to reconstruct images on a screen. By rotating an x-ray beam around a patient or other specimen in three directions, CAT can produce a two-dimensional picture on a screen. See tomography; microtomography.

cedarwood oil. An essential oil from the "cedar" tree _Juniperus virginiana._ It is processed by clearing and evaporation, and is useful in microscopy as a medium for oil immersion lenses. At the temperature at which it is used for this purpose its refractive index should be 1.515. Nowadays immersion liquids (q.v.) are man-made and customized for particular oil-immersion objectives.

central stop (for condenser). An opaque disk placed in the ring carrier or diaphram carrier of the substage apparatus. It excludes the central rays and is used for darkfield work at low magnification. <u>Variable stops</u> can be contracted or expanded as desired. For <u>colored stops</u> see optical staining; Rheinberg disks.

central stop (for objective). The opaque stop usually placed in the objective back focal plane to give dispersion staining.

changing devices. The mechanical contrivance on the objective end of the microscope tube, for mounting the objective. The usual device is the well-known rotating nosepiece (*q.v.*) which may carry from two to six objectives, a later development of which makes it possible to center each objective separately. Individual centering rings are also available for the rotator without the centering feature.

chromatic aberration. See aberration, chromatic.

chromatic light filter. See color filter; optical light filters.

circle of confusion. The image of a point in object space in the image plane, sometimes called the antipoint. An immeasurably small object point will be represented by a disk in the image plane, even by a perfect lens but its size and shape is altered by lens defects. An image can be considered as a composite of many such disks. When the primary image is to be enlarged, the original circle of confusion should be so small that it will still appear sharp after enlargement. As the lens aperture decreases, the diffraction component of the circle of confusion will increase in size and, with a good lens, this should dominate. In a photographic image photographic factors can affect the size and fuzziness of the circle of confusion.

cleavage. The property of a crystalline substance of splitting along definite crystal planes.

collimation. Controlling a beam of radiation so that its rays are as nearly parallel as possible.

collimating lens. A lens used to produce a collimated beam from a light source. With a true point source on the axis the beam would be an axial tube of parallel rays. With practical sources having real area the beam will be a diverging cone of parallel rays from each light point; this beam still has special properties and can be collected again by a good lens.

colloid. A suspension of sub-light-microscopic particles. This definition arbitrarily limits the size of the particles to 0.1 - 0.005 mm. Such particles may be studied by darkfield illumination, particularly with the light ultramicroscope or by means of an electron microscope.

colophony. A natural resin, the rosin of commerce. It may be used in microscopy as a mounting medium. It has an index of refraction of about 1.545. It is sometimes called colophonium. See resin; rosin; resinography.

color filter. The more common term for an optical chromatic filter. Any filter which produces a chromatic effect when placed in the path of the illuminating rays.

color stops. See optical staining; Rheinberg filter.

color temperature. A reference term for expressing the quality of the light of an incandescent light source that depends upon the fact that most of such sources exhibit the same light quality (spectral distribution) as does a black material when incandescent although at different temperatures. The quality of the light from any incandescent source is expressed as the temperature of a black body when it emits the same quality of light; it is expressed in degrees Kelvin (°K) of the absolute temperature scale. The concept is so useful that it is often used when not strictly applicable. As the temperature is raised light quality changes from a dull red towards a blue-white. The color temperature of an incandescent tungsten lamp is usually 2800° - 3000°K. The color temperature of "daylight" (sun in a blue sky) has been internationally defined (ANSI) as 5500°K.

coma. A lens aberration which occurs in that part of the image field which is slightly away from the principal axis of the system. It results from different magnifications in the various lens zones. It causes extra-axial object points to appear as short comet-line images, with the tail either toward the center of the field (positive coma) or away from the center (negative coma). Coma is fundamentally due to the faulty position of the principal points (*q.v.*) of the lens.

comparison standard. A standard micrograph or a series of micrographs, usually taken at 75 or 100 diameters, or a suitable equivalent built into the eyepiece and used to determine, *e.g.*, grain size by direct comparison with the image.

compensating ocular. A correcting eyepiece of either the positive (Ramsden, q.v.) or negative (Huygens, q.v.) type. The negative types are usually used for low-power oculars and the positive types for high-power ones. The corrections are of a high order and they are intended for use with apochromatic objectives of all powers. The residual under-correction for color associated with apochromats is compensated for by the over-correction of the compensating ocular (hence its name). For this reason a ring of yellow appears around the image of the ocular diaphram in the field of view.

compensator. A device for determining the degree of retardation (*q.v.*) and hence the degree of birefringence (*q.v.*) in an anisotropic specimen. The most common compensator is the first-order red plate (full-wave plate) but a $1/4\lambda$ plate is also available. A quartz wedge (*q.v.*) is a common type of variable compensator. The Babinet compensator employs opposing quartz wedges. The Berek compensator tilts an anisotropic material.

complementary colors. Two colors which, if combined, would yield white or gray.

compound lens. A lens combination of two or more simple lenses. A corrected lens such as a microscopical objective, composed of two to twelve or more elements, is therefore a single compound lens.

compound microscope. See microscope, compound.

conchoidal fracture. Type of fracture seen when a mineral, such as quartz, or a glass, breaks to give irregularly curved and usually striated surfaces.

condenser. A term applied to lenses or mirrors designed to collect, control, and concentrate radiation in an illuminating system. See condenser (substage).

condenser, Abbe. See Abbe condenser.

condenser circle. The image of the aperture iris diaphram of the substage condenser (_q.v._) as seen in the back focal plane of the objective.

condenser (substage). In microscopy, the lens mounted before the microscope stage, which transmits light to the object. There are two main categories of condensers: (1) brightfield and (2) darkfield. Brightfield condensers are of four distinct types: (a) Abbe, (q.v.), an uncorrected condenser composed of two separable lenses; (b) aplanatic condenser; (c) achromatic condenser which has full corrections for color and spherical aberration; (d) aplanatic-achromatic condenser. The darkfield condenser for low powers may be nothing more than a low-power brightfield condenser with a central stop. Medium- or high-power darkfield condensers are usually of the cardioid (_q.v._) or paraboloid (_q.v._) type. The lamp lens is loosely called a condenser lens, but light-collecting lens is a more definite term. All microscope condensers must be carefully focused and aligned for best results.

condenser, darkfield. A condenser forming a hollow cone of light with its apex (or focal point) in the plane of the specimen. Used with an objective having a numerical aperture lower than the minimum numerical aperture of the hollow outer zones from the achromatic condenser. The best condenser for critical visual work and photomicrography.

condenser, variable-focus. Essentially an Abbe condenser in which the upper lens element is fixed and the lower movable. The lower lens may be used to focus the illumination between the elements so that it emerges from the stationary lens as a large diameter parallel bundle. The field of low-power objectives may thus be filled without removing the top element of the condenser. At the opposite extreme it can be adjusted to have a numerical aperture as high as 1.3.

cone of light. The light cone formed by the microscope condenser. A full cone is obtained when the focused condenser is adjusted to just fill the objective aperture with light. A half cone, a 9/10 cone, etc., is obtained when the diameter of the condenser circle (*q.v.*) is 1/2, 9/10, etc. the diameter of the objective circle (*q.v.*).

conjugate foci. In an image-forming system, two fields are said to be conjugate with each other when two or more object fields are simultaneously in focus in a single plane, *e.g.*, in Köhler illumination the field diaphram, specimen and ocular front focal plane are conjugate foci as are the filament, condenser aperture diaphram, objective back focal plane and ocular front focal plane (when a Bertrand lens is used).

conoscopic observation. The study of the back focal plane of the objective by removing the eyepiece, by inserting a Bertrand lens, by examining the image at the eyepoint above the eyepiece with a magnifier or by using a phase telescope is called conoscopic because the observations are associated with the cone of light furnished by the condenser and viewed by the objective (*cf.*, orthoscopic).

contrast micrometer. A micrometer scale composed of alternate black and white squares which replace the lines in an ocular disk micrometer. The additional contrast makes measuring easier and faster on many subjects.

Corex® glass. A glass, made by the Corning Glass Company, quite transparent to ultraviolet radiation. It is used for microscopical slides and coverslips intended for fluorescence work. In a measure it replaces the more expensive quartz. Corex® glass is also used as a filter glass. It should not be subjected to contact with water.

cornea. The comparatively hard covering on the outside of the front part of the eyeball.
correction collar (objective). A screw adjustment, generally under constant spring tension, which affords a method for controlling under- or over-correction by altering the spacing of the two rear lens combinations in the specific 2 mm and 4 mm objectives. It

makes possible the adaptation of the lens to various coverslip thicknesses, usually from 0.10 to 0.20 mm. The graduated collar is turned until the optimum image is obtained, as determined by the star test (*q.v.*) (which employs a pin hole to act as a "star") or, more often, by critical examination of the image.

correction of lens. The image by a simple lens is degraded by various aberrations such as chromatic (*q.v.*), spherical (*q.v.*) as well as curvature of field, coma (*q.v.*) and astigmatism (*q.v.*). Each must be corrected for use in any optical instrument.

correction of an objective. A microcope objective corrected for one wavelength (spherically) and two wavelengths (chromatically) is said to be achromatic. An apochromatic objective is corrected for three wavelengths (chromatically) and two (spherically). A semi-apochromat, often called a "fluorite" corrections intermediate between the chromat and apochromat.

crossection. A thin section of a solid substance prepared for microscopical viewing by microtomy. (See microtome).

counting ocular. An ocular with a reticle projecting a network of squares within a prescribed boundary upon the field. After calibrating the size of the squares with a stage micrometer (*q.v.*), a count can be made of the number of particles, or any other items, per unit area of specimen. (Comment: A reticle of concentric boundaries, allowing an estimate of size is not a counting ocular.)

coverslip. Thin glass, plastic, sapphire, SiO_2, NaCl, etc., shaped into circles, squares, or rectangles for covering the specimen. Glass coverslips, grade no. 1 1/2, contain a high proportion of covers from 0.17 to 0.18 mm in thickness, close to the lens designer's specifications for employment of achromatic and apochromatic objectives of high NA. The thickness and refractive index of the glass coverslip are important because certain of their values have been assumed by the designer and manufacturer. These values have been specified by ASTM: thickness, commercial number 1 1/2: 0.16 to 0.19 mm; refractive index measured at the sodium D line (589.3 nm), 1.523 ± 0.005.

critical angle. The maximum angle of incidence that can be formed by a light ray in passing from a medium of high refractive index to one of lower index. The angle is measured between a ray and a perpendicular erected at the intersection of the ray with the surface of the medium. The sine of this angle is the reciprocal of the refractive index of the medium. It is the angle of total reflection.

crown glass. Most ordinary window glass is crown glass. Alkali-lime glass used in lenses is also crown. It is known as a soft glass and has low dispersive power. An achromatic doublet is composed of two lens elements, a positive lens made of crown glass and a negative lens made of flint glass (_q.v._) having higher dispersion.

crystal, birefringent. A crystalline (_q.v._) substance which is anisotropic (_q.v._) with respect to the velocity of light.

crystalline. A substance (usually solid but can be liquid) in which the atoms, ions, or molecules are arranged in a definite pattern that is repeated regularly in three dimensions. Solid crystals tend to develop forms bounded by definitely oriented plane surfaces that are harmonious with their internal structure. They may belong to any of six crystal systems: isometric, ("cubic"), tetragonal, hexagonal, orthorhombic, monoclinic, or triclinic.

"crystalline" lens of the eye. The small lens which focuses light rays on the retina (_q.v._). The normal lens has a variable focal length of 14 to 20.7 mm (after Helmholtz's schematic eye). The change in focal length is attained by muscular effort which causes alteration in the radius of curvature of the lens' surfaces, thus producing accommodation for far and near points.

cubic. Having the shape of a cube. In crystallography the cube is the chief form (morphological shape, _i.e._, "habit") in the isometric system. In microscopy the term "cubic" often appears also to designate the isometric, isotropic system of crystal structure.

curvature of field. The image plane formed by a single lens is naturally curved. While one part of the field will be in good focus, the rest will need refocusing to be sharp. While the eye may partially correct for this, a camera lens will not, and the final image as photographed will not be in perfect focus over the entire image plane.

dammar. A natural resin obtained from evergreen trees in the East Indies, especially from Agathis australis. It is closely allied to the kauri gum or copal of the paint manufacturers. In microscopy it is used like Canada balsam (*q.v.*) for mounts. It is also used to seal mounts.

darkfield condenser. See condensers. The ordinary brightfield condenser of low power, used with a central stop, makes a good darkfield condenser. See special darkfield condensers: paraboloid, cardioid and Cassegrainian. They all form a dark field while illuminating the specimen with a hollow cone of light. The lower limiting aperture of the condenser must be greater than the NA of the objective with which it is to be used. Thus, no direct light enters the objective; the specimen is seen by reflected or scattered light on a dark background.

darkfield illumination. Any method of illumination which illuminates the specimen but does not admit light directly to the objective. It may be by substage (darkfield, *q.v.*) condensers; by stage spot lighting, by special condensers fitted around special objectives for reflected illumination or by the slit ultra-microscope.

darkfield objective. Certain objectives for high-power, darkfield work equipped with iris diaphrams or funnel stops so that their apertures may be reduced to correspond to the darkfield condenser with which they are used.

darkfield slides. Owing to the exacting demands of darkfield illumination, not only must the microscope slide be especially clean, but also the glass of which the slide is composed must be optically clear under darkfield conditions. The glass should not fluoresce. For general use and for special work, see ASTM specifications.

darkfield stop. A central stop for obtaining a darkfield effect for low-power objectives. It is customarily used with a high NA, brightfield condenser.

Davis shutter. A fitting with a small iris diaphram, attached above a low-power objective for reducing the aperture. In this way depth of field can be increased for the photomicrography of objects illuminated by incident light.

daylight, artificial. Artificial daylight may be obtained by placing the microscope and lamp near a window and examining the field of view alternately with daylight and lamplight. Optical filters are added until the lamplight simulates a daylight effect. Results obtained in this way may be considerably different from those obtained by the stereotype method of merely inserting a "daylight filter glass."

dendrite (dendritic). A branched tree-like crystal habit usually associated with rapid crystal growth, usually polycrystalline (_cf._ skeletal).

definition. The distinctness with which the very fine detail in an image or photograph can be seen.

densitometer. A photometer to measure point by point the density of exposed photographic film.

density. A term applied particularly to the rating of neutral filters (see opacity) and to the measurement of the density of negatives. Mathematically, it is the logarithm (base 10) of the reciprocal of the transmission.

depth of field. The depth (thickness) in the object space that is simultaneously in acceptable focus.

depth of focus. The depth (thickness) in the image space that is simultaneously in acceptable focus.

diaphram (optical). The deliberate outer limit to the size of a beam of light entering or leaving a lens or lens system. The diaphram used to restrict the effective size of a lens should lie in its principal plane or be imaged in it. In a lens system, such as a microscope with its lamp, an aperture diaphram is one that is imaged in the objective aperture and limits the angular extent of the illumination beam. The field diaphram is imaged in the microscope field of view and protects the image from "stray light". In a photographic lens the diaphram is sometimes called a "stop".

diffraction. When light passes an edge, because wave undulation is involved, it tends to curl around the edge, as would water waves. This small fraction is usually not observed as it is overwhelmed by the rest of the undeviated light. However, when two parallel edges are brought close together to form a slit, excluding most of the unaffected light, this spreading of the light according to wavelength becomes apparent; the shorter the wavelength the more it is bent—blue more than red. With many slits together, *i.e.* a grating, this is most evident and it can be used instead of a prism. When there are two parallel slits another phenomenon can be observed, i.e. interference (*q.v.*). Abbe showed that it is diffraction at the minute edges of the specimen that limits observable detail with the microscope. In this case the interference involved is with the undeviated beam of light.

diffraction disk. See Airy disk.

diffraction grating. An artificially produced periodic array of scattering centers capable of producing a pattern of diffracted energy, such as accurately ruled lines on a plane surface.

diopter. An optical unit representing the reciprocal of the focal length (in meters) of a lens, in terms of the meter. A 1-diopter lens has a focal length of 1 meter, a 2-diopter lens has a focal length of 1/2 meter, etc. The diopter expresses the power as in spectacles or other weak lenses.

dioptrics. An old term, relating to study of image formation by a lens due to refraction. It is not included in the general term, geometrical optics.

dispersion. The variation of refractive index with color (or wavelength) of light. The spreading of white light into its component colors when passing through a glass prism is due to dispersion which, in turn, is due to the fact that the refractive index of transparent substances is lower for long wavelengths than for short wavelengths. A measure of dispersion ν is defined as:

$$v = \frac{n_D - 1}{n_F - n_C}$$

where n_D = refractive index at 589 nm (yellow); n_F = 486 nm (blue); n_C = 656 nm (red).

dispersion staining. A procedure involving central or annular stops in the objective back focal plane to induce colored images of transparent particles mounted in liquids with indices matching the particle at a wavelength in the visible. The particle and liquid should possess very different dispersion curves for best colors.

distance of virtual image. When a simple lens is used as a magnifier for visual observation the eye becomes part of the optical system. A "virtual image" can be formulated by construction and its apparent distance will vary with the focus of the eye. This will vary among individuals. In a rather arbitrary but standardized assumption, the "normal distance" for close observation, or reading has been set at 10 inches (250 mm). Therefore, by agreement, the magnification of a simple magnifier:

$$m = \frac{10\ (inches)}{focal\ length\ (inches)} = \frac{250\ (mm)}{f.\,l.\ (mm)}.$$

The optics for the compound microscope have been designed to furnish parallel light from the eyepiece so that the eyes are relaxed for distant viewing. This makes the virtual image lie at "infinity". Tests showed that the average observer accommodates somewhat, placing the virtual image rather variably, often about 20 - 25 feet.

distortion. The aberration of image distortion results when the magnification varies over the extent of the field.

(1) barrel distortion. This distortion results when the magnification increases from the center toward the edges of the field. This is also called negative distortion.

(2) pincushion distortion. This distortion is the result of decreasing magnification outward from the center of the field. It is also called positive distortion.

dominant wavelength. Any color can be matched by specifying a monochromatic (spectral) wavelength, and the amount of white light that must be mixed with it to match the specification or sample, except for some colors, such as purple, where two wavelengths are required. In the latter case the wavelength, of the complementary color is used to keep the simple definition. The character of the white light (*e.g.*, daylight) must be specified. This system is used to specify the color of optical filters that are used visually. It is less useful for color filters to be used photographically but is sometimes used anyway.

double refraction. The refraction of light in two slightly different directions to form two rays or vector components. Each ray is polarized, and their vibration directions are perpendicular to each other. Furthermore, each ray has a different velocity, and therefore a different refractive index. See birefringence.

doublet. Two simple lens elements cemented together, for the purpose of giving spherical and chromatic corrections. An achromatic doublet may be composed of a positive crown-glass lens and a negative flint-glass lens. The errors of one element are made to compensate those of the other element.

drawing prism. The prism of a camera lucida (*q.v.*) which permits both the field of view and the drawing surface to be examined simultaneously. Also a reversing prism placed at the exit pupil of the microscope to direct the image-forming rays onto a drawing board or vertical screen.

drawtube. The smaller of the two tubes on a monocular microscope. The drawtube (if present) carries the ocular; it can be adjusted to control tubelength and so effect corrections for the objective lens. The drawtube may also be convenient in calibrating an eyepiece-micrometer by allowing a minor magnification change to simplify the conversion factor.

electromagnetic radiation. Propagated forms of energy with electrical and magnetic fields and with wavelengths from gamma rays through x-rays, ultraviolet, visible, infrared to radio waves.

electromagnetic lens. An electromagnet designed to produce a suitably shaped magnetic field for the focusing and deflection of electrons or other charged particles in electron optical instruments (*cf.* electrostatic lens.)

electron beam. A stream of electrons in an electron optical system.

electron diffraction. The phenomenon, or the technique of producing diffraction patterns through the incidence of electrons upon matter.

electron gun. A device for producing and accelerating a beam of electrons.

electron image. A representation of an object formed by a beam of electrons focused by an electron optical system (see image).

electron lens. A device for focusing an electron beam.

electron micrograph. A photographic reproduction of an image formed by the action of an electron beam.

electron microscope. See microscope, electron.

electron microscopy. The study of materials by means of an electron microscope.

electron optical axis. The path of an electron through an electron optical system along which it suffers no deflection due to lens fields. This axis does not necessarily coincide with the mechanical axis of the system.

electron optical system. A combination of parts capable of producing and controlling a beam of electrons to produce an image of an object.

electron optics. The science that deals with the propagation of electrons, as light optics deal with that of light and its phenomena.

electron probe. A narrow beam of electrons used to scan or illuminate an object or screen.

electron-probe microanalyzer (EMA). The qualitative and quantitative use of x-rays excited by a microprobe of electrons. Available with SEM (*q.v.*) and TEM (*q.v.*).

electron trajectory. The path of an electron.

electron velocity. The rate of motion of an electron.

electron wavelength. The wavelength necessary to account for the deviation of electron rays in crystals by wave diffraction theory. It is numerically equal to the quotient of Planck's constant divided by the electron momentum, mv.

$$\lambda = \frac{h}{mv}$$

where: h = Planck's constant
m = mass of electron
v = velocity of electron

electrostatic lens. A lens employing a permanent magnet to produce a potential field capable of deflecting electron rays to form an image of an object. (cf. electromagnetic lens).

electrostatic spraying. A technique of spraying wherein the material being sprayed is given a high electrical charge, while the specimen is grounded.

entrance pupil. The apparent size of the limiting aperture of a lens or lens system (properly that of the diaphram), as seen from the object plane. This can shift and become a complex matter in some circumstances. In a properly set up microscope system it should be that of the substage iris diaphram.

epsilon. (ε). Greek letter representing the "extraordinary", unique refractive index in a uniaxial (*q.v.*) crystal. See omega (ω).

equant. A shape having nearly equal dimensions.

equivalent focal length (e.f.). The focal length of the simple lens that has the same power as the compound lens. The stated focal lengths of microscope objectives are the e.f. since their front and back focal lengths are very different.

etching. Controlled preferential attack on a surface for the purpose of revealing structural and morphological details.

euhedral crystals. Those that are bounded by plane faces; (_cf._, anhedral).

eutectic. The lowest melting mixture of two (or more) substances miscible in the melt state.

eutectic arrest. In a cooling curve (or heating curve) of two or more components, an approximately isothermal segment, corresponding to the time interval during which the heat of crystallization is being evolved, (or conversely).

eutectic, equilibrium (eutectic). The composition within any system of two or more crystalline phases which melts completely at the minimum temperature, or the temperature at which such a composition melts.

eutectic point. The composition of a liquid phase that is in univariant equilibrium with two or more solid phases; the lowest melting composition.

eutectic structure. The construction resulting when a solution has passed through a eutectic equilibrium upon freezing.

exit pupil. The exit pupil of a lens system is an image of the entrance pupil (hence conjugate to it) and normally should be the image of the limiting diaphragm. In both the microscope and the telescope it is the eyepoint where the beam has its smallest crossection. It is also called the Ramsden circle (_q.v._) or eyepoint.

exposure index. The speed rating of a photographic film for use in connection with exposure tables, exposure computers, and exposure meters.

exposure scale. In a photographic process, the range of exposures over which substantially correct reproduction is obtained. This is measured by the ratio of the exposure corresponding to the minimum useful gradient at the high exposure end of the scale to that corresponding to the minimum useful gradient at the low exposure end.

exposure test. An exposure made as part of a process to establish the amount of incident radiation that will produce a satisfactory image.

extinction. While using a polariscope, such as a polarized light microscope, extinction of the light image occurs when a birefringent object (such as a crystal or fiber) is turned between crossed polars to such a position that its illumination on the dark field has been extinguished. Usually extinction of the birefringent object occurs at 90° rotation intervals of the stage, (which is perpendicular to the axis of the microscope or other polariscope). See polariscope, polars, polarized light, etc.

extinction, oblique. Vibration directions oblique to the long direction of the crystal or fiber.

extinction, parallel. Vibration directions parallel and perpendicular to the long direction of the crystal or fiber.

extinction, symmetrical. Vibration directions bisecting a prominent crystal profile angle.

extinction angle. The angle between the nearer vibration direction and a prominent direction of the crystal. It never exceeds 45°.

eye lens. The lens nearest the eye in any ocular.

exposure time. The time during which the receiving surface is irradiated (*e.g.*, photographic film).

eyepiece. The lens system near the eye which magnifies the primary image of the objective; ocular.

eyepiece, parfocal. Oculars with common focal planes so that they are interchangeable without substantial refocusing.

eyepiece, negative. An ocular in which the real image of the object is formed between two lenses. The Huygens type (*q.v.*).

eyepiece, positive. An ocular in which the real image of the object is formed below the lower lens of the eyepiece. The Ramsden type of eyepiece (*q.v.*).

eyepoint. See exit pupil.

F-number. A measure of the relative aperture of a lens; its light gathering capacity. It is equal to the ratio of the focal length of the lens divided by the diameter of its limiting opening (aperture): f-number = focal length/iris diameter.

Note that the number becomes smaller as the aperture grows larger and that it must be squared to directly measure the area which is the light gathering capacity. Also see numerical aperture.

false Becke line. A second bright line which moves in the direction opposite to the Becke line. It is usually observed with thick particles or when the refractive index difference between particles and mountant is large. It appears as a concentration of reflected light at the high index side of the interface.

far point of eye. For the normal eye, the far point is at infinity. The rays of light from an infinitely distant point source are parallel and can be focused with the accommodation muscles of the eye entirely relaxed. See accommodation; near point of the eye.

fast ray. The fast ray or fast component for a crystal or fiber corresponds to the lower refractive index.

field. The area of the object that is seen when the image is observed. It may range in diameter from several millimeters to less than 0.1 mm. The magnification and the size of the diaphram opening in the eyepiece governs the diameter of the field of view.

field depth. The thickness of the object space within which objects focused by a lens will all appear in good simultaneous focus. Penetration is a synonym.

field diaphram. In a photomicrographic system particularly, an iris diaphram that is imaged in the field of view with Köhler illumination. This limits the extent of the illuminated field and eliminates much extraneous light. See diaphram.

field-emission microscope. Either one of two kinds of point-projection microscopes, both invented by E. W. Muller:

(1) The older device (1936) is a specialized cathode-ray tube, employing field-emission of electrons from a negatively charged tip of a very sharp needle in a vacuum, by point-projection of the image onto a positively charged, fluorescent screen.

(2) A later device (field-ion-emission microscope, 1950) emits absorbed helium ions from an anode.

field lens. The lower lens in an ocular; the lens nearest the object field.

field of view. The extent of the visible image field that can be seen.

filament. An electrically heated wire used as a source of radiations, such as electrons, or as a source of heat such as for the vaporization of a metal.

filar micrometer ocular. A micrometer ocular with crossed lines which can be simultaneously focused in and moved across the field by means of a micrometer screw. The amount of displacement can be read in 0.01 steps on the micrometer drum head. This makes a very delicate measuring device, since calibration of the micrometer scale by means of a stage micrometer gives values for each interval on the drum head which are much less than the resolving power of the highest-aperture objectives.

film cassette. See cassette.

film speed. A measure of image intensity for a film relative to a given intensity of illumination.

filter. A device which modifies radiation coming from the source with regard either to wavelength or intensity.
(1) _color filter_. A screen inserted into a light beam to absorb part of the spectrum. The ideal color filter would transmit one wavelength efficiently and absorb all others. Interference filters approach this ideal. The next best filter for maximum transmission combined with sharpness of cut-off is a

liquid containing organic dyes or certain inorganic salts, followed by Wratten dyed gelatin filters and the glass filters containing certain metals and metallic salts. Color filters help to eliminate chromatic aberrations. They also increase contrast in the image, and green filters permit more restful observation.

(2) _contrast_. A color filter, usually with strong absorption, whose function is to utilize the spectral absorption bands of the subject to control the contrast of the image by exaggerating or diminishing the brightness difference between areas of different color. Maximum contrast is obtained when the transmission of the filter is entirely within the absorption band of an area but not of its surroundings.

(3) _interference filter_. A combination of several thin optical films to form a layered coating for transmitting or reflecting a narrow band of wavelengths by virtue of interference effects.

(4) _neutral_. (a) A color filter that reduces the intensity of the transmitted illumination without affecting its hue; (b) a color filter having identical transmission at all wavelengths throughout the spectrum. Such an ideal filter can only be approached in practice.

(5) _orthochromatic_. A color filter whose function is to modify the illumination quality reaching photographic film so that the brightness of colored objects will be relatively the same in the resultant black-and-white positive.

(6) _photometric_. A color filter whose function is to convert the quality of illumination from that of one source to that of another.

(7) _x-ray filter_. A material that either preferentially absorbs x-rays or transmits them.

(8) _ultraviolet filter_. A material that either absorbs or transmits ultraviolet rays. Its use is required in some aspects of fluorescence microscopy.

fine adjustment. Mechanism needed for precise focusing. One full turn of the fine-adjustment knob of a light microscope is usually equivalent to a vertical motion of the microscope tube of 0.1 mm. Thus, if the knob is engraved with 100 equally spaced marks, the sensitivity of the adjustment will be 1 μm.

fine motion. See fine adjustment.

fixing bath. See hypo.

flint glass. See glass, flint.

fluorochromes. Dyes which have little coloring effect under ordinary lighting conditions but which fluoresce when irradiated with ultraviolet radiation.

fluorescence. The absorption of radiation to produce radiation of longer wavelengths, *i.e.*, lower energy.

fluorescence microscope. A microscope illuminated by ultraviolet or blue light so that the object may re-radiate light of longer wavelengths. To protect the eyes, a UV-absorbing filter should be provided if not built into the fluorescence microscope.

fluorite. A mineral (crystalline CaF_2) used in lens manufacture. It belongs to the isometric (cubic) crystal system and has the low refractive index of 1.434.

flux, light. Sometimes called luminous flux; the visible portion of the radiant energy emitted by a light source. It is measured in lumens per solid angle. In electrical engineering, it is analogous to the lines of force in a magnetic field, spoken of as magnetic flux.

focal length. The distance from the optical center of the lens to the focal point. The focal length of an objective and its working distance (*q.v.*) are directly proportional.

focal point. The point on the principal axis of a lens at which impinging parallel rays converge to a common point or focus. The point is virtual for a negative lens. See virtual image.

focus. A point at which rays, originating from a point in the object, converge.

focusing device (electrons). A device which effectively increases the angular aperture of the electron beam illuminating the object, rendering the focus more critical.

focusing glass (light microscopy). A hand magnifier, used at the focal plane of a camera, usually with the ground glass removed. Its purpose is to examine the image for critical focus. It is, as a rule, a low-power magnifier so mounted that it is in focus when supported on a piece of clear glass at the focal plane of the camera, for studying the image. It operates equally well on the aerial image.

focusing magnifier (electron microscopy). A low-power light microscope, telescope, or simple lens used to observe the electron image formed on a fluorescent screen.

focus, principal. The point at which a lens focuses an infinitely distant axial object point. Synonymous with focal point (*q.v.*).

fog (of a photographic material). The blackening due to other than the recording of the image upon exposure.

Note: - fog may be caused by the following: accidental exposure of the emulsion to light or penetrating radiations, stray radiation, improper environment of the emulsion before use (including chemical, mechanical and thermal), aging of the emulsion, and improper processing.

foot-candle. A measure of illumination. The illumination received at a surface 1 foot from a standard lamp of 1 candlepower; 1 foot-candle = 1 lumen per square foot.

form. In crystallography, a group of faces such as a cube, octahedron, etc. (closed forms), or prism, basal pinacoid, etc. (open forms). In this sense, the term form is not to be confused with phase (*q.v.*) in physical chemistry.

Formvar™. Trademark for polyvinyl formal resin, a plastic material used in preparing replicas or specimen-supporting films, especially in electron microscopy. Trademark registered by the Monsanto Co.

free aperture. The effective diameter of the free aperture of a diaphram or lens. It is the diameter of the unrestricted surface between the edges of the mount of the lens, or of the unrestricted area of the diaphram. Telescope objectives are rated according to the diameters of their free apertures.

frequency. With reference to electromagnetic wave motion, the number of complete waves, or cycles, per second.

Fresnel, Augustin Jean (1788-1827). A French mathematician and optician. Designer of the lens (*q.v.*) which bears his name.

Fresnel fringes. A class of diffraction fringes formed when the source of illumination and viewing screen are at a finite distance from a diffracting edge. In the electron microscope these fringes are best seen when the object is slightly out of focus.

Fresnel lens. A lens built up, progressively, in zones or steps, each zone with its own individual radius. Considerable spherical correction is attained, and the weight of the lens is greatly reduced. Fresnel lenses were originally designed for lighthouses but they are now attainable for small spotlights, automobile headlights, and similar uses.

Fresnel reflection. Process by which radiant flux is reflected from an optically plane boundary between two transparent dielectric materials.

front lens. The front element of a compound lens system; the first lens element which the entering light encounters. (See back lens.)

gamma. The contrast of a negative or a positive controlled by developing time, or otherwise. Numerically, it is the tangent of the angle of the straight-line portion of the D (density) *vs.* log E (exposure) curve.

gamma (γ). Greek letter representing the highest principal refractive index in a biaxial (*q.v.*) crystal or the angle between the a and b axes of a triclinic crystal.

Gauss points. The points on the optical axis of a lens through which the principal planes (*q.v.*) pass. (Named after Karl Friedrich Gauss).

glare. Stray or scattered light within the microscopical system; extraneous light from windows or lamps; light scattered by the specimen that masks image detail and prevents thorough observation. An under- or over-corrected objective or condenser also may cause glare.

glass, inorganic. A fused, amorphous mixture, the usual constituents being silica (SiO_2) and oxides of a variety of elements (chiefly Na, K, Ca and Pb, sometimes Mg, B, Zn, Al, and others), in supercooled solution of such high viscosity that at ordinary temperatures it is, essentially, solid, hard, brittle and usually transparent. The constituents of a particular glass determine its refractive index, light dispersion and other properties. Examples of glasses important to microscopy include crown glass (*q.v.*) and flint glass (*q.v.*).

glass, crown. 75-85% silica (usually fine white sand) by weight plus 10-20% alkali oxides which consist mostly of sodium oxide (from soda ash), 5-10% calcium oxide (from lime), potassium oxide (from potash) or magnesium oxide (from carbonates). Optically, conventional crown glass has a relatively low index of refraction and small dispersion.

glass, flint. Although most constituents of flint glass are similar to those of crown, a substantial amount of lead oxide ("red lead") replaces the calcium oxide in crown. Flint glass has a higher refractive index and strong dispersion, resulting in the "brilliance" of "lead" glass.

glass, optical. Meeting at least the following requirements: (a) no noticeable birefringence (by adequate annealing); (b) highest homogeneity; and (c) highly transparent to some portion(s) of the spectrum (usually the visible).

glass, organic. Certain polymeric, organic thermoplastics, optically important for their chemical composition, transparency, clarity, and low refractive index; for example, poly(methyl methacrylate) (PMMA), known commercially as Acrylite™, Lucite™, Plexiglas™, etc.

glass, silica. An almost pure SiO_2 glass, i.e., vitreous silica. It is extremely difficult to manufacture due to the high temperatures required (other glasses have fluxes such as sodium oxide added to reduce the requisite working temperature), its very high viscosity (making it difficult to "work"), and changes in refractive index from even slight impurities within the glass. Transmits ultraviolet light.

glassy state. A solid state, anhedral in shape and with conchoidal fracture; if free from strain, isotropic. These materials may change to the crystalline state, _i.e._, devitrify, in time by the action of temperature, shock, scratching, etc. Glassy materials do not melt at a specific temperature but soften gradually and continuously on heating.

goniometer. An instrument for measuring the interfacial angles of a crystal.

goniometer ocular. A special type of ocular fitted with crosslines and a graduated head which rotates the crosslines to facilitate the measuring of angles, as in crystallography.

goniophotometer. Photometer for measuring the directional light distribution characteristic of sources, lighting fittings, media and surfaces. Also called a distribution photometer.

grating, diffraction. A series of narrow, close, equally spaced, diffracting slits or grooves capable of dispersing light into its spectrum. Diffraction gratings and their replicas are also used as standards in micrometry, especially in electron microscopy.

grating, reflection. An opaque (metallic) diffraction grating from which incident light is reflected to form a spectrum, or to act as a micrometric standard for opaque specimens.

36

grating, transmission. A transparent diffraction grating.

Greenough microscope. One of two kinds of stereomicroscopes (_q.v._) with two separate compound microscopes, one for each eye, focused on the same object. The other kind has a common main objective (CMO). See also binocular microscope.

ground glass focusing screen. A light-scattering glass pane, one side of which has been ground or otherwise made diffusing, mounted in a camera in lieu of photosensitive material for the purpose of intercepting, viewing, focusing and measuring a real image formed on the ground glass. To return a small area to a clear glass focusing screen (_q.v._), a sheet of coverslip large enough to receive a magnifier is cemented to the light-scattering side.

habit. The morphology (shape) of a crystal. In each of the six structural systems, the same terms are appropriate to describe the particular crystalline habit: lamellar, tabular, equant, columnar, or acicular, etc.

haemacytometer. A glass slide with a chamber for counting blood corpuscles in a given volume. The cells are 0.1 mm deep and are made with several types of ruling. Such cells are convenient counting chambers for many other kinds of suspensions. Also spelled hemacytometer; hemocytometer.

hanging drop slide. A glass slide with a concavity or a built-up chamber which allows a drop of culture to be placed on a coverslip inverted over the cell. It makes possible the examination of freely moving protozoa, etc., confined only by the limits of the drop and the bottom surface of the coverslip.

heat-absorbing glass. A glass, useful as a filter for microscopy and photomicrography. It absorbs principally in the infrared which carries most of the heat of the spectrum. Since the glass becomes warm and reradiates, it is wise to put it in a water-filled cell. Because of its composition, the glass is brittle and easily broken. Such a filter is the alternative to an infrared reflecting glass filter.

hemimorphic. A crystal with unlike faces at opposite ends of the same axis.

hemisymmetric. A crystal on which one-half the faces of at least one form are systematically missing.

holoscopic eyepiece. An eyepiece suitable for achromatic, fluorite and apochromatic objectives. The eyelens can be adjusted closer to or further away from the field lens to spherically correct for the various types of objectives. (This adjustment also changes the magnification of the eyepiece.)

homal. A device made by Zeiss to replace the eyepiece of a microscope for photomicrography. Since it consists of a negative lens that lies below the plane of the primary image from the objective, it projects this objective image into the final image plane, correcting its defects as would the eyepiece. Since there is no eyepoint (it is virtual) it cannot be used visually. Their advantage was a somewhat flatter field. (They appeared before the era of modern "flat-field objectives.")

homogeneous immersion objective. An objective to be immersed in a liquid of a certain refractive index and dispersion value as specified by the manufacturer of the objective. An oil-immersion objective (*q.v.*), the most important type, is intended to be immersed in "cedarwood" oil (n_D = 1.515) or in its manmade optical equivalent. A water-immersion objective is for dipping into an aqueous specimen mount. Alpha-monobromonaphthalene (*q.v.*) has such a high refractive index (n_D = 1.66) that a very highly resolving objective (1.60 NA) was designed to be immersed in that liquid, for use by reflected light on metals and other opaque objects.

Huygens or Huyghens, Christian (1629-1695). Dutch mathematician, astronomer, natural philosopher, inventor of the Huygenian ocular, and originator of the wave theory of light. His diagrammatic scheme of showing the advance of wavefronts, and how each wavefront consists of points each considered as individual original sources for the development of successive wavefronts, is still in general use.

Huygenian eyepiece. An undercorrected (blue rim at periphery of field) ocular designed by Huygens for the telescope and later adopted for achromatic objectives. This ocular consists of two plano-convex lenses separated by a diaphram, hence it is of the negative type with the focal plane inside the system. Both plane surfaces face the eye, as distinguished from the Ramsden eyepiece (*q.v.*).

hyperfocal distance. The nearest distance, h, at which a camera lens focuses the far distance (infinity). This gives the maximum depth of field; the nearest distance in focus is h/2.

hyperopia. A condition in which visual images come to a focus behind the retina of the eye; farsightedness. No eyeglasses are required while using a microscope (unless there is excessive astigmatism also). See astigmatism; myopia.

hyperplane eyepiece. A special eyepiece made by Bausch & Lomb for flattening the visual field.

hypo. A solution of sodium thiosulphate, $(Na_2S_2O_3 \cdot 5H_2O)$, sometimes called sodium hyposulphite, used to dissolve the silver salts of photosensitive material which have not been affected by light. It is used after the negative has been developed. See fixing bath.

Iceland spar. Calcite, mineral, $CaCD_3$, of optical quality.

illumination, critical. The Nelsonian method in which the light source is imaged in the plane of the specimen. A ribbon filament or arc lamp is required to give uniform illumination; the lamp must be focusable; the filament position must be adjustable in all directions. The use of an achromatic (*q.v.*) condenser is advised.

illumination, Köhler. This method images the field diaphram in the plane of the specimen and the light source itself in the back focal plane of the objective.

illumination, oblique. Illumination from light inclined at an oblique angle to the optical axis.

image. A representation of an object produced by means of radiation, usually by means of a lens or mirror system.

image, aerial. A real image formed on a screen or any surface becomes an aerial image when the screen is removed. It actually exists in space, as can be proved by blowing smoke across the space it occupies.

image field. Any field showing a focused image. There are a number of such fields in the complete microscopical system. The term may also denote the field of view, or the image field at the focal plane of the camera, generally the field where the final image is formed.

image, real. An image as formed by a lens on a screen, plate or any plane surface. See image, aerial; image, virtual.

image space. The space about an optical system each point of which is conjugate to some point in the object space (*q.v.*).

image, virtual. A virtual image has no real existence. It is the image seen when looking into a mirror. The field of view of the microscope is a good example of a virtual image. When the eye operates in conjunction with a lens to form an image on the retina, the visual sensation is as if the image existed in space. That its apparent location is very definite is proved when a pin can be made to coincide with the mirror (virtual) image of another pin that is seen by looking at a sheet of glass acting as a mirror. With a lens system a virtual image can be definitely located as by graphically tracing rays back to a focus. In a microscope, if the eye is relaxed as it should be, the virtual image will be at infinity. Measurements show that most observers place the aerial image at 20-25 feet, some as close as seven, because of partial accommodation.

immersion liquid. Any liquid occupying the space between the object and microscope objective. Such a liquid is usually required by objectives of 3-mm focal length or less. For best results (*i.e.*, resolution) the liquid should be used between the condenser and the microscope slide. Immersion objectives for transmitted light are designed for use with either "oil", glycerine, _or_ water, the refractive index of the liquid and the coverslip (if any) being the determining factor. The liquid and the front lens of the objective should ideally coincide in index and in dispersion value. See oil immersion; homogeneous immersion of objective.

immersion of a lens. With nearly all high-power lenses, it is intended that the spaces between the condenser and the slide, and the specimen and the front lens of the objective be filled with an immersion liquid. Owing to the limitations imposed by the critical-angle phenomenon, numerical apertures are impossible exceeding 1.0 without immersion. In addition, immersion makes possible the use of the naturally aplanatic points of the front lens element of the objective.

immersion objective. An objective in which a medium of higher refractive index than that of air is used in the object space to increase the numerical aperture and hence the resolving power of the lens. See homogeneous immersion of objective.

immersion objective (electron optics). A lens system in which the object space is at a potential different from that of the image space.

incidence, angle of. If a perpendicular is erected at the point of incidence of a light ray, at a surface, the angle between the perpendicular and the light ray is the angle of incidence. See plane of incidence.

incident light (illumination). Sometimes denotes any over-stage lighting not included by methods of vertical illumination.

inclusion. Fluid (liquid and/or gas) droplets, and/or solid particles often incorporated in a solid matrix, especially a crystal.

index of refraction. The ratio of the velocity of light in a vacuum to its velocity through a transparent or translucent substance. For optical materials that velocity is always less than in a vacuum. Thus, for this class of material the index is always greater than 1.

indicatrix. A three-dimensional construction of the optical relationships in a crystal. Radii proportional in length to either refractive index or velocity may be used to represent variations in those values throughout a crystal (or oriented polymer fibers and films).

infrared radiation. For photomicrography the energy radiation extending beyond the long visible red rays of the spectrum. The image is invisible to the eye. Thus it must be studied by photographic recording or by image translators.

intensity. Intensity of a light source is measured by its output of luminous flux, in terms of lumens per solid angle. It is reduced to candlepower by dividing by 4π. Intensity of a source is also spoken of in terms of brightness (candles) per unit area (see brightness). The latter term usually applies to sources with extended surfaces.

interfacial angle. That angle between two faces of a crystal observed along their zone axis. The angle between the perpendiculars to two crystal faces is termed the polar angle.

interference. Two light beams which originate from the same source and are identical spectrally and in polarization are capable of interference (_i.e._, are coherent) because they are a wave action.

(1) If they are still in phase or one has been retarded by $n\lambda$ they will add completely to each other's brightness. This is called <u>constructive</u> <u>interference</u>.

(2) If one beam has been retarded, with respect to the other, by $n\lambda/2$, _i.e._, from crest to trough, the two waves nullify each other to the limit of the weaker beam, causing darkness if the two are equal; this is <u>destructive</u> <u>interference</u>. There is only partial interference when the retardation of one beam is only intermediate between $n\lambda$ and $n\lambda/2$.

Since complete destructive interference occurs strictly for a single wavelength, independently, it results in residual color when white light is involved.

42

interference colors. See polarization colors; Michel-Lévy color chart of birefringence/retardation *vs.* thickness.

interference figure. The conoscopic pattern of extinction positions of a crystal superimposed on the pattern of interference colors corresponding to the full cone of directions by which the crystal is illuminated, each direction showing its own interference color.

intermediary plane. When in an optical system several successive real image planes of an object are formed, reticles or diaphrams can be inserted at an "intermediate plane" for superposition on the final image.

interpupillary distance. The distance between the centers of the pupils of one's eyes.

inverted microscope. A microscope so arranged that the line of sight is directed upward through the objective to the object.

iris diaphram. A diaphram (*q.v.*) with thin metal leaves so arranged that the rotation of an actuating lever varies the diameter of the opening.

isogyres. In a uniaxial interference figure, the two black brushes that form a cross and represent the pattern of extinction positions of the crystal. In a biaxial interference figure, the two black (usually curved) brushes that intersect the isochromatic curves (lemniscates) and represent the pattern of extinction positions of the crystal.

isometric system. Embraces crystals with three mutually perpendicular axes having identical spacings of atoms. Isometric crystals (also termed cubic crystals) are consequently isotropic (*q.v.*) unless strained.

isomorphous. Having the same crystal structure and similar morphology (habit) but different chemical composition yet forming a continuous series (system) of solid solutions; occurs when the molecular shapes and sizes are closely similar.

isotropic. Substances having no polarizing effects on light and therefore remaining dark between crossed polars. Crystals in the isometric ("cubic") system are consequently isotropic. Such a substance has only one value of refractive index, irrespective of the vibration direction of the light passing through it (*c.f.*, birefringence).

isotropy. The condition of having the same values for a vector property in all of its directions.

Köhler illumination. A highly effective illumination system for reflected-light as well as transmitted light microscopy. The image of the field diaphram is focused on the specimen surface, and the image of an undiffused lamp source is focused in the plane of the aperture diaphram.

Lagrange disk. The exit pupil of the microscope; also called exit pupil, eyepoint and Ramsden circle or disk.

lambda (λ). A symbol for wavelength of light, usually expressed in nanometers (nm).

lambda zero (λ_0). In dispersion staining (*q.v.*), the wavelength at which both specimen particle and immersion liquid have the same refractive index.

lambert. A photometric unit for describing the brightness (*q.v.*) of a surface. One lambert is the equivalent of 1 lumen per square centimeter.

lamellae. Thin, flat scales or plates in parallel array.

lapping. Rubbing two surfaces together, with or without abrasives, for the purpose of obtaining dimensional accuracy or superior surface finish.

law of constancy of interfacial angles. In all crystals of the same substance (in the absence of polymorphism), angles between corresponding faces are identical.

law of rational indices. The lengths of intercepts of different faces of a given crystal on any crystallographic axis are in ratios of small integers.

lens. An optical element, so constructed that it serves to change the degree of convergence or divergence of the transmitted rays.

lens, Bertrand. A small convergent lens, which, when placed between objective and eyepiece, focuses an image of the back focal plane of the objective on the front focal plane of the eyepiece. It is chiefly used with polarized light for inspecting the interference figure. It is also convenient for quickly focusing and centering the filament for Köhler illumination, and to verify centering and size of the field diaphram.

lens, compound. A lens composed of two or more separate pieces of glass or other optical material. These component pieces are usually cemented together. A common type of compound lens is a two-element magnifier, one element being a converging lens of crown glass and the other a diverging lens of flint glass. The combination of suitable glasses or other optical materials (plastics, minerals, etc.) properly ground and polished reduces aberrations always present in a single lens.

lens, eye. The lens (in an eyepiece) nearest to the eye.

lens, negative. A lens that is thicker on the edges than in the center, and which causes parallel light rays to diverge. Also called a diverging lens or amplifier (*q.v.*).

lens, simple. A glass disk ground and polished with a spherical figure on one side and a plano, concave, or convex configuration on the other side.

light. Light is radiant energy of such wavelength that, falling on the retina, it stimulates the rods and cones of the eye and produces the sensation of vision. The foregoing is a physiological (subjective) definition which tells what light does, under certain conditions, rather than what it is. According to the (objective) Maxwell theory, all radiant energy is electromagnetic (*q.v.*) in character, the generation of the radiation depending upon the portion of the spectrum under consideration. For that part of the spectrum lying within the visual range, approximately 400 to

700 nanometers (nm) the release of light energy from externally stimulated sources may be thought of as due to atomic or molecular vibration or to the passing of electrons from high to lower energy levels accompanied by the spasmodic release of energy as the electronic orbits decrease in diameter. See quantum theory.

light filters. See optical light filters.

light flux. See flux, light.

light-gathering capacity. The amount of light flux gathered by a lens system. Microscopical objectives have light-gathering capacity proportional to the squares of the numerical apertures of the objectives.

light microscope. A microscope (*q.v.*) employing the visible or near-visible portion of the electromagnetic spectrum.

light sources. The main light sources available in microscopy are:
      (a) tungsten-filament lamps, the most useful of which are the projection lamp and the ribbon-filament lamp;
      (b) capillary gaseous discharge tubes of the mercury-vapor type, such as the H series of the General Electric Company;
      (c) arcs - carbon, xenon, or argon;
      (d) quartz-halogen.

      Natural daylight is unsuitable for microscopy and especially for photomicrography.

longitudinal magnification. A certain distance, measured axially, in the object space as referred to the respective distance in the image space. The ratio of the distance D' in the image space to the distance D in the object space equated to the square of the linear magnification M of the system. Thus: $M^2 = D'/D$.

LM. Abbreviation for light microscope (*q.v.*) or light microscopy (*q.v.*).

lumen. A measure of light flux. A source of candlepower emits 4π lumens; 1 foot-candle = 1 lumen per square foot; 1 lumen =

0.0015 watt, $\lambda = 555$ nm.

luminance. The luminous intensity of any surface in a given direction per unit of projected area of the surface as viewed from that direction.

luminous flux. See flux, light.

luminous spot ring condenser. See Cassegrainian darkfield condenser.

macrograph. A close-up photograph at a low magnification up to 40X or thereabouts.

macroscopy. The interpretive use of the eye, generally aided by a hand lens up to 10 or 20X in magnification.

magnification. The number of times by which the size of the image exceeds the original object. Lateral magnification is usually meant. It is the ratio of the distance between two points in the image to the distance between the two corresponding points in the object.

magnification, empty. Any magnification, whether obtained by the microscope or by subsequent enlargement, which is greater than the minimum requirement to show the resolvable detail. Conventionally it is magnification greater than 1000X NA.

marker, object. A small abrasive stylus, set in a rotating holder mounted on the lower end of the drawtube. The desired part of the specimen is placed in the center of the field, and the abrasive point is pressed against the slide or cover, and rotated. It describes a tiny circle around the desired object field.

maximum. As used in diffraction and interference phenomena to denote those parts of the diffraction pattern where the light energy is most concentrated.

mechanical stage. A device provided for adjusting the position of a specimen by translation in two directions at right angles to each other.

mechanical tubelength. Measurement from the shoulder of the objective to the upper end of the drawtube. The generally accepted length for most microscopes is now 160 mm. Metallographic objectives and some others are corrected for use with longer tube lengths -- 190 to 215 mm or more. See correction of lenses; correction of an objective; optical tubelength.

melatope. The center of rotation of the isogyres in biaxial interference figures representing the point of emergence of rays that, in the crystal, travel along the optic axes.

melting point. In a phase diagram, the temperature at which the liquid and the last trace of solid co-exist at an invariant point, i.e., an equilibrium melting or freezing point.

mercury-vapor discharge tube. The mercury-vapor lamp. Several types are manufactured by the General Electric Company. Those known as capillary discharge tubes, or the H series, are bestsuited for microscopical illumination. They are known as cold discharge lamps; the current is passed between the electrodes by gaseous conduction. Traveling electrons raise the mercury-vapor atom electrons to higher energy levels (energy is absorbed), and, on the return of the electron to its normal or lower level, light energy is emitted.

mercury-vapor lamp. See mercury-vapor discharge tube.

metallography. That branch of science relating the constitution and structure to the properties of metals and alloys.

Michel-Lévy scale of retardation colors. Color chart plotting thickness of the anisotropic specimen, its birefringence (n_1-n_2) and its retardation in nanometers. Any one of the three variables can be determined if the other two are known.

microbiological contaminants. Living organisms imvisible to the naked eye that can contaminate a specimen.

microcut. The purposeful scratch made by a microhardness tester.

micrograph. A graphic reproduction of the image of an object as formed by any microscope.

microhardness test by microindentation. A hardness test which employs a calibrated apparatus to force a diamond indenter of specific geometry, under a test load of 1 to 1000 g, into the prepared surface of the test material in order to measure the diagonal or diagonals microscopically as a measure of hardness.

microhardness test by scratching. A hardness test which employs a corner of a diamond under a standard weight to make a scratch the width of which is then measured microscopically.

micromanipulator. A mechanical device for making small movements in order to manipulate or treat a microscopic object with a fine needle, small forceps, micropipette, or other very small tool.

micrometer (μm). Systeme Internationale (SI) unit of one millionth of a meter; 0.001 mm; formerly called a micron (μ). Micrometer ('mi-cro-meter) is not to be confused with a measuring device called a micrometer (mi-'crom-eter). See micrometer ocular; filar micrometer ocular.

micrometer disk. A glass disk engraved with a suitable scale, used at the diaphram of a micrometer ocular. The scale to be focused by the eye lens has to be seen in the field of view.

micrometer eyepiece. See micrometer ocular; micrometer disk.

micrometer ocular. An eyepiece (with a focusable eye lens) which carries a micrometer scale. The linear scale must be calibrated by means of a stage micrometer scale for each objective with which it is used.
micron (μ). Obsolete. See micrometer.

microphotograph. A small, microscopic photograph, in which the image is minified; it requires enlarging or the use of a lens system in order to view it. See photomicrograph.

microphotometer. A photometric device to fit onto the microscope tube. It measures the intensity of light reflected or transmitted from the specimen, for estimating optimum exposure in photomicrography (*q.v.*).

microscope. An optical instrument capable of producing a magnified image of a small object by means of particles, sound waves or electromagnetic radiation.

microscope, compound. An optical instrument wherein a primary image is formed by the first lens, or objective, and focused by a second lens to form a virtual image with the aid of the eye, or a real image by projection. Both compound light and electron microscopes usually employ a third lens, the condenser to control illumination. See also microscope, simple.

microscope, electron. An electron-optical device which produces a magnified image of an object. Detail may be revealed by virtue of selective transmission, reflection, or emission of electrons by the object.

microscope, field emission. An image-forming device in which a strong electrostatic field causes cold emission of electrons or ions from a sharply rounded point or from a specimen that has been placed on that point. The electrons or ions travel to a phosphorescent screen, or photographic film, giving a visible picture of the variation of emission over the specimen surface. See field emission microscope.

microscope, stereoscopic. Either one of two kinds: binocular-binobjective, such as the Greenough (*q.v.*) type, and binocular microscope (*q.v.*) with common main objective (CMO). See stereomicroscope.

microscope, simple. A single lens system, sometimes a doublet or triplet, which acts as a magnifier.

microscope mirror. Usually plane on one side and concave on the other. The flat side is generally used unless the objective is of very low power and there is no condenser. The mirror should be so mounted that the concave side can be focused on the specimen.

50

microscope, x-ray. A device for producing enlarged images of a specimen by means of x-rays. Dioptric systems, analogous to light microscopes, are not available, but contact microradiography, point-projection, and reflection techniques provide practical alternatives.

microscopic. Very small, pertaining to a very small object or to its fine structure. A microscopic particle requires microscopical examination to be adequately visible.

microscopical. Pertaining to a microscope; pertaining to the use of a microscope.

microscopy. The interpretive use of microscopes.

microspectroscope. A spectroscopic eyepiece which makes possible the analysis of light transmitted by or reflected from a specimen. The better spectroscopic oculars are fitted with a wavelength scale, and one is in the form of a direct-vision spectroscope.

microstructure. The structure or texture of a suitably prepared specimen as revealed by a microscope. See structure; morphology.

microtome. A device for cutting thin-sections of relatively soft materials for examination by transmitted light or electron microscopy, It cuts successive thin slices using a knife of metal, glass or diamond.

miller indices. The notations usually used for naming crystal faces; they have the general formula hkl. These notations are based on the assignment of crystallographic axes and on an expression of the intercepts of the face on the three axes (hexagonal has four, hikl).

minification. A reduction of the lens image from the true size of the object.

minimum. Part of the diffraction or interference pattern where destructive interference has caused darkness. The dark rings surrounding the central disk of a diffraction pattern are minima. See diffraction.

monobjective binocular microscope. A microscope with one objective and two bodies, for binocular vision, not necessarily stereoscopic.

monochromatic light. Light composed of one wavelength. It may be obtained by the use of a laser or by gaseous discharge tubes in combination with proper filters. An approximation is obtained by interference filters or monochromators.

monochromatic light filter. An optical light filter passing only a very narrow spectral band. It is best used with a gaseous discharge lamp where the energy sources are concentrated into a few narrow bands of very restricted wavelengths.

monochromatic objective. An objective, usually made of fused quartz, which has been corrected for use only with one visible light wavelength.

monochromator. A light-dispersing instrument which is used to obtain light of substantially one wavelength, or at least of a very narrow band of the spectrum. The light is apt to be weak because of its purity. The use of monochromatic light filters is usually adequate for microscopical work.

monocular microscope. A microscope with one objective and one bodytube for monocular vision.

monomer. A chemical entity capable of reacting with itself over and over again (to form a homopolymer) or with another monomer (to form a copolymer). Certain monomers and polymers are useful in microscopy for mounting specimens.

morphological analysis. The identification of particles based on easily observed microscopical characteristics, such as size, shape, and organization of parts. See morphology.

morphology. The size and shape of an animate or inanimate structure, its parts, or particles; topography of a surface; habit of a crystal; distribution of phases in a system. Compare *structure*.

mounting medium. Any liquid, polymer, resin, melt, or even gas used to mount microscopical specimens before examination. The chief attributes of a mounting medium are refractive index, viscosity, lack of color and long-term chemical stability.

mounting melt. A solid solution of elements or compounds. For use as a mounting medium, the components are purposely of high index, *e.g.*, sulphur and selenium. The medium must be melted for use and remain isotropic at room temperature.

mu. The Greek letter μ. It is the symbol of one-millionth part, as in micrometer μm (*q.v.*).

muscae volitantes ("flying flies"). Dead cells floating in the vitreous humor of the eye. They are sometimes troublesome in monocular microscopical vision because they cast shadows on the retina. However, their disagreeable optical effects can be largely eliminated by substituting binocular vision, or by using a larger aperture and higher power ocular.

myopia, nearsightedness. A condition of the eye in which the rays from distant objects are brought to a focus before reaching the retina. Eyeglasses not required while using a microscope because focusing the microscope accommodates the eye.

NA. Abbreviation for numerical aperture (*q.v.*).

nanometer (nm). 10^{-9} meter or 10A. However, A (angstrom, *q.v.*) is not an SI unit. See *SI*.

near point of the eye. The nearest point to the entrance of the pupil of the normal eye at which focus is attained without strain; 10 inches (250 mm) is the generally accepted distance. In very young people and in those with myopia, this distance is less. The near point recedes with age, possibly even causing farsighted condition, presbyopia (*q.v.*) or hyperopia (*q.v.*).

negative crystal. A uniaxial crystal is optically negative if $\varepsilon<\omega$. A biaxial crystal is said to be optically negative if $\gamma - \beta<\beta - \alpha$.

Otherwise the crystal is positive.

negative eyepiece. An eyepiece in which the real image of the object is formed between the lens elements of the eyepiece. See negative ocular; ocular.

negative focal length. The focal length of a negative lens. Parallel rays impinging on a negative lens can be traced to a virtual focus which exists on the same side of the lens as the impinging rays. The distance from the second principal point of the lens to this second focal point is measured on the same side of the lens as is the object. See negative lens; focal length.

negative image. A developed photographic image in which dark areas correspond to light areas of the original subject. See negative, photographic.

negative lens. A lens with a negative focal length. The edge of a negative lens is thicker than the center. The three negative lenses are, according to their figure: planoconcave; double concave or biconcave; and diverging concavoconvex or diverging meniscus. See lens, simple.

negative ocular. A Huygenian ocular or one based on the Huygenian principle. See ocular.

negative, photographic. A sensitized plate or film which has been exposed in a camera and which, upon development, has the lights and shades reversed to those of the original subject. The plate or film does not become a negative until it is exposed, after which it may be an undeveloped or a developed negative. See positive, photographic.

negative print. A photograph (opaque or transparent) having approximately the opposite rendition of light and shade to the original subject.

Newton's interference colors. Newton's series of colors by interference results when two wave-trains of white light meet. For some wavelengths (colors) there will be destructive interference (darkness) but for other wavelengths there will be reinforcement (color). They are most frequently observed in very thin films, only wavelengths thick, *e.g.*, oil on water. The interference is caused by partially reflected light from the interfaces. Newton's series of colors appears in sequence in the Michel-Lévy chart (*q.v.*).

nicol prism. A polarizing element made of two pieces of calcite specially cut, ground, polished and cemented. A transmitted beam splits into two polarized components, one of which is refracted into and absorbed by the asphalt mount. The remaining polarized beam is transmitted. Modifications of this prism are common. The Nicol prism ("nicol") has generally been superseded by a polarizing film such as Polaroid™. See polars; polarizer; analyzer. ™ signifies registered trademark of the Polaroid Corporation.

normal. An imaginary line forming right angles with a surface or other lines; sometimes called the "perpendicular". It is used as a basis for determining angles of incidence, reflection, and refraction.

nosepiece, rotating. A device on the end of the microscope tube to permit the mounting of two to six objectives, any of which may be swung into place, ready for use, by rotating the nosepiece to the desired position. The nosepiece usually occupies a mechanical-tubelength space of 15 mm.

numerical aperture (NA). The numerical aperture of a lens system (objective or condenser). It is the sine of one-half the angular aperture times the refractive index of the medium (1.0 for air, 1.515 for Cargille immersion oil, etc.) between objective and specimen. The numerical aperture is a measure of the light gathering capacity of the lens system and determines its resolving power and depth of field.

object field. A position lying in the front focal plane of the objective.

object space. A space within which an object could be imaged by the lens.

objective. In any optical instrument, the lens system nearest to the object. High-powered light microscopical objectives are of very complicated construction and computation, often containing ten or more separate lens elements. The lower element, or front lens, affords the magnification and is of high power. The subsequent lens elements serve to bend the image-forming rays back to the lens axis at the ocular front focal plane and to correct, insofar as possible, the aberrations of the objective. The front lens is usually a single lens element; the back lenses may be doublets or triplets according to the intention of the lens computer.

objective circle. When the focused objective is examined by withdrawing the ocular and viewing its back focal plane through the microscope bodytube, the limiting boundary of the objective, the circumference of the white spot of light, is the objective circle.

objective, fluorite. An objective using the mineral fluorite in its construction. It is usually intermediate between achromatic and apochromatic in correction.

obtuse bisectrix(Bx_o). The angle supplementary to the acute optic axial angle of a biaxial crystal is obtuse and is bisected by the obtuse bisectrix, which may be either α or γ.

ocular. Essentially, nothing more than a hand magnifier. It serves to magnify a primary image. It operates with the eye to form a virtual image, or, in photomicrography it forms a real image of the primary microscopical image, the difference in operation being attained by simply focusing the microscope to bring the primary image into the proper position. The two main types of oculars are the Huygenian (_q.v._) and Ramsden (_q.v._) types. They can be obtained with varying degrees of correction. They also appear as compensating oculars. Huygenian oculars are of the negative and Ramsden oculars of the positive type. (See negative ocular; positive ocular.) Amplifying lenses (see amplifiers) used in place of oculars are suitable only for projection or for photomicrography. Note: An ocular serves very well as a true hand magnifier when removed from the microscope and inverted.

ocular micrometer. A glass disk which can be introduced into standard oculars, upon one surface of which a fine scale is accurately engraved. See filar micrometer ocular; reticle.

oil immersion. See immersion liquid.

omega (ω). Greek letter representing the "ordinary" refractive index a uniaxial crystal. See epsilon (ε).

opacity. Opacity of a transparent material is the reciprocal of its light-transmitting value. See density, optical.

optic axis. Of a doubly refractive crystal, that direction along which no double refraction occurs. Compare axis, optical.

optic sign. For uniaxial materials, by definition, the sign is positive if the value $\varepsilon-\omega$ is positive, and negative if $\varepsilon-\omega$ is negative. For biaxial materials, the sign is positive if γ is the acute bisectrix, and negative if α is the acute bisectrix, or positive if $\gamma-\beta > \beta-\alpha$ and negative if $\gamma-\beta < \beta-\alpha$.

optical chromatic filter. See color filter.

optical density. Thickness of a transparent object times its thickness along a chosen light path.

optical flat. Usually, a glass or quartz plate or disk, the thickness of which should be at least 1/10 of its diameter. It is ground until any remaining unevenness can be measured only by interferometric methods. Their maximum departure from flatness usually is less than 1/10 of the sodium doublet (589.3 nm).

optical glass. An especially fine glass made under the most carefully controlled conditions. There are many kinds, some with low index and high dispersion values and some with high index and low dispersion. Good lenses are always made from optical glass.

optical index. A constant applied to objectives for purposes of comparison. It takes into account the focal length or magnifying power of the lens and also the NA. It was offered by Nelson and by Coles, but little use has been made of the optical index figure, probably because the equations of the two men differed and the resulting figures have been at variance.

optical light filters. Filters made to modify the light source either chromatically (chromatic filters), in intensity (neutral filters), or thermally (heat-absorbing filters). They may be of solid glass, stained gelatin sheets mounted or unmounted between glass plates or glass cells filled with an appropriate liquid.

optical microscope. A very ambiguous term since all microscopes involve optics; better to specify light, acoustic, x-ray or electron microscope, etc.

optical staining. Producing color in the microscopical image so as to differentiate one part of the object from another. One way is by use of Rheinberg filters (_q.v._). Another is to use polarized light (_q.v._) on an anisotropic (_q.v._) specimen. Another important method is by dispersion staining (_q.v._).

optical tubelength. The distance from the second principal focal plane of the objective to the front focal plane of the ocular. This distance is about 180 mm for high-power objectives, less for those of low power.

orthoscopic observation. The normal way of viewing an object microscopically (cf., conoscopic observation). With Köhler illumination the field diaphram and the ocular front focal plane as well as the specimen will be in simultaneous focus.

over-correction. A lens that focuses central light rays nearer to the lens than it focuses outer rays. The reverse is under-correction.

paraboloid darkfield condenser. A lens of parabolic shape. The vertex end is ground back so that its focus can be brought into coincidence with the specimen on the slide. A central stop is provided to block the central rays. It is used chiefly for medium-power work.

parallel rays. A beam of light from a point source, on the axis and at the focus of a lens, emerges from the lens as a beam of parallel rays, i.e., a tube of rays within the diameter of a lens. With practical sources, made up of a composite of points, the emerging beam is an expanding cone made up of parallel rays of light from each point. This is true of the emerging beam from a microscope ocular when properly set up for visual use.

paraxial. Referring to the spacing of rays closely surrounding the principal axis of a lens system.

parfocal objectives. Objectives which are mounted so that only a small adjustment of the bodytube and stage is necessary to focus after changing from one objective to any of the others. They are mounted in such a way that the back focal plane is in the same position on the optical axis of the microscope for each objective. Objectives used on a rotating nosepiece are usually parfocal. Eyepieces are also parfocal within any given manufacturer's series.

pedesis. See Brownian movement.

pedion. In crystallography, a single face having no equivalent.

pellicle. A thin film or skin that reflects some of the incident light and transmits the rest. It functions as a beam-splitter.

penetration. See field depth.

perception. The awareness, and usually the meaning of a stimulus, or stimuli, based on previous experience.

periplan eyepiece. A compensating type of eyepiece made by Leitz with a doublet for the eyelens. The field is larger and flatter than the regular compensating ocular, but there is apt to be spherical aberration or distortion present near the periphery of the field.

permanent magnet lens. An electron lens consisting of permanent magnets.

Petzval condition. This is based on a theory which states that the sum of the product of the refractive indices and focal lengths of two thin lenses must equal zero in order to attain a fairly flat field free from astigmatic conditions.

phase. In physical chemistry, a homogeneous, separate portion of a material system. For example, calcite is phase I of $CaCO_3$; aragonite is phase II, and vaterite is phase III. Each phase has its characteristic crystallographic, optical, and other physical properties. Also: solid and liquid phases of the same composition; polymorph, eutectic, addition compound, or solid solution phase in a composition diagram.

phase angle. The angular equivalent of the time displacement between corresponding points on two sine waves of the same frequency.

phase-amplitude contrast. The separation and recombination of direct vs. diffracted rays in a light microscope adjusted to Köhler illumination (*q.v.*). At the lower focal plane of the condenser there is an annular diaphram with an opaque central stop. Through this diaphram rays are focused as a hollow cone onto the specimen. In the back focal plane of the objective there is a conjugate annular diaphram (phase plate). If here the undiffracted rays are retarded (by a transparent film of proper thickness on the annulus of the phase plate), bright contrast results. If, instead, the phase-delay film is on the central spot, dark contrast results. With either a bright or a dark-contrast phase plate, the annulus is usually coated with a partially absorbing (very thin) film of silver (Zernike method) or carbon soot (Wilska method) to reduce the higher amplitude (intensity) of the undiffracted rays.

phase (optics). In any oscillatory motion, such as light, the position and character of a wave at any instant, often with respect to its zero or maximum position.

phase shift. A change in the phase relationship between two alternating quantities of the same frequency.

photomacrograph. Photograph in the magnification range of conventional macrophotography (<~40X) taken with a single lens system, often a microscope objective.

photomacrographic lens. A photographic lens system (often a microscope objective) specially mounted for close-up objects and used without an ocular.

photomacrography. Close-up photography at a limiting magnification of 40X or thereabouts and taken with a single lens system such as a photomacrographic lens (*q.v.*) or a microscope objective.

photomicrograph. An image enlarges approximately 40X or higher, produced by light, *cf.*, electron micrograph.

photomicrography. This term should not be reversed into microphotography. A photomicrograph is a photograph of a small object, the image of which is magnified more than approximately 40X by means of a compound microscope. A microphotograph is a small photograph, requiring an enlargement or a lens system in order to view it; the image is minified.

photosensitive material. Usually, films or plates sensitive to light action; also refers to printing paper or any surface treated to produce a photographic effect.

pinacoid. Two parallel faces of a crystal having no equivalents.

planoscopic eyepiece. An American Optical Company (Reichert) eyepiece designed to flatten the field of achromatic objectives.

plane, focal. A plane through the focal point ideally perpendicular to the principal axis of a lens or mirror.

plane glass illuminator. For brightfield reflected illumination, *e.g.* for metallography, a thin flat disk located just above the back lens of the objective and acting as a partial mirror to direct illumination to and from the specimen. It allows complete use of the objective aperture which an opaque mirror partially occludes. See vertical illumination.

plane of incidence. The plane containing a ray incident to a surface, and the normal to that surface, erected at the point of incidence.

plane wave. Wave in which wavefronts are parallel to a plane normal to the direction of propagation.

plano. In optics, an optical surface which has been made substantially flat, the degree of flatness depending upon the performance required. A plano-convex lens is a positive lens with one surface flat and the other convex. In the dictionary, plano is given only as a combining form, but in practical optics it is often used alone to denote any particularly flat surface—one that has been worked flat. See optical flat.

pleochroism. The phenomenon of a substance showing different absorption colors in different vibration directions of polarized light. The observed colors change with the orientation of the crystal and can be seen only with plane polarized light, *i.e.*, a single polar.

polar (objective) as in polar angle. The angle between two perpendiculars to two lines (or planes) forming a true angle; as a noun it refers to polarizing filters, see polars.

polar interfacial angle. The angle between the perpendiculars to two crystal faces.

polarimeter. An instrument fitted with polarizer (*q.v.*) and analyzer (*q.v.*) for detecting the direction and extent of rotation of polarized light by an aqueous solution of an optically active substance such as the sugars.

polariscope. An instrument, such as a polarized light microscope (*q.v.*), for analyzing the effect of a specimen on polarized light. See polarizer; analyzer; polars.

polarization colors. Interference colors produced by anisotropic (*q.v.*) substances placed between two polarizing elements and examined by transmitted white light. See Michel-Lévy scale of retardation colors.

polarized light. Light that is vibrating in one plane (plane-polarized light), light with a rotary vibration (circular polarized light), or light that is vibrating elliptically (elliptically polarized light). Moonlight and skylight are polarized, as is much reflected light; cloud light is polarized under certain conditions. However, naturally polarized light is, on the whole, rather imperfectly polarized.

polarizer. A polarizing element (polar) which is placed between the light source and the preparation with its vibration direction preferably set in an E-W direction. See polars.

polarizing element. A polar; term for a device for producing or analyzing plane polarized light. It may be a nicol prism, some other type of calcite prism, a reflecting surface, or a polarizing film (such as made by the Polaroid Corp). See polars.

polarized light microscope. A microscopical polariscope, i.e., a compound microscope which is equipped with two polars (*q.v.*) and a Bertrand lens (*q.v.*); chemists and mineralogists are the principal users.

polarizing prism. See nicol prism.

Polaroid®. The trademark of the Polaroid Corporation, used in reference to its camera; photosensitive materials, polarizing film, etc. Like all trademarks, it must be capitalized, designated by ® or ™ and defined. Example: Polaroid® polarizing film.

polars. Two polarizing elements in a polarized light microscope. The polar placed between the light source and substage condenser is called the polarizer; the polar placed between the objective and ocular is called the analyzer. The vibration directions of the two polars may be crossed 90° to achieve "crossed polars"; slightly uncrossing one polar gives "slightly uncrossed polars"; removing the analyzer results in "plane-polarized light".

polymer. A macromolecular substance characterized by the numerous repetitions of monomeric units joined together by chemical bonds. See monomer.

polymorphism. The phenomenon of crystallizing in two or more different space lattices, resulting in different crystal structures, and usually different kinds (forms) of crystal faces. Thus, for a given chemical element or compound, polymorphism results in two or more crystalline phases, each with its own specific set of physical properties.

positive crystal. A uniaxial crystal is optically positive if $\varepsilon>\omega$. A biaxial crystal is said to be optically positive if $\gamma-\beta>\beta-\alpha$. Otherwise, the crystal is negative.

positive focal length. Any lens which converges parallel rays to a focus at the back of the lens is a positive lens (*q.v.*) and has a positive focal length. See negative focal length. The focal length is measured from the second principal point of the lens to the point on the lens axis where the rays from an infinitely distant point are brought to focus.

positive image. A developed image in which dark areas correspond to dark areas of the original subject.

positive lens. Any lens with a positive focal length. Such lenses arc thicker in the center than around the circumference. There are three types of positive lenses; double convex or biconvex, planoconvex, converging concavoconvex or converging meniscus.

positive ocular. A Ramsden ocular or any modification of it. The diaphram is below the field lens (*q.v.*).

positive, photographic. A photograph having approximately the same rendition of light and shade as the original subject.

positive transparency. A photographic positive made on a transparent base. In electron microscopy this is used as an intermediate step to prepare a negative print.

preferred (or principal) direction. Orientation of a specimen as related to a structural or morphological direction, with respect to the polars.

presbyopia. Farsightedness due to age; inability to focus nearby objects. See hyperopia (far-sightedness); cf, myopia.

64

principal direction. See preferred direction.

principal plane. A plane normal to the principal axis of a lens, passing through a principal point (*q.v.*).

principal point. Mutually conjugate points lying on the lens axis and producing a magnification of one. Object and image distance as well as focal length are calculated (measured) from these points.

prism (prismatic). Crystal made up of three, four, six, eight or twelve similar faces all parallel to a single axis.

prism (optical). A transparent body with at least two polished plane faces inclined with respect to each other, from which light is reflected or through which light is refracted. When light is refracted by a prism whose refractive index exceeds that of the surrounding medium, it is deviated or bent toward the thicker part of the prism.

prism illuminator. A 45 to 90-degree prism interposed in the tube of a light microscope for the purpose of directing an intense beam of light through the objective onto the object. The prism illuminator utilizes about one half the aperture of the objective as does the mirror illuminator (*q.v.*) with an attending loss of resolution over that obtainable with either a plane glass illuminator or a pellicle mirror. See pellicle; metallography.

prism, nicol. See nicol prism.

profilometer. An instrument for measuring the profile of a surface by moving a stylus over the surface and recording its calibrated amplified motion.

profile angle. An angle, not necessarily an interfacial angle, used to describe a crystal. This angle is observed when the crystal is lying on a face. For example, a cube shows 90°; an octahedron 60° or 120°.

projection lamps. Light bulbs to fit projectors of images onto a retroreflective screen (*q.v.*). A projection lamp usually employs a coil filament that may not give a uniform field of illumination in a microscope. A ribbon filament is much better. Diffusing the light, as by frosted glass, is not good microscopical practice.

projection ocular. Strictly, low-power oculars which are designed for projection purposes at distances of 16 feet or more. Homal (*q.v.*) lenses are not classified as projection lenses but as amplifiers (*q.v.*).

projection lens (electron). The final lens in the electron microscope. The lens focuses the electrons onto a fluorescent screen to produce the visible image.

pyramid (pyramidal). A group of three, four, six, eight or twelve similar faces intersecting in a point or parallel to faces that would intersect in a point.

quantum theory. According to this theory, after Max Planck, radiant energy is given off from atoms or molecules in small discrete units called quanta. It is absorbed in a like manner. The Bohr conception of the atom in relation to its emissivity ties in the dis charge of the energy with the passing of an electron from a high energy level to a lower level (from a larger to a smaller orbit), and the passage of an electron from a smaller to a greater orbit is accompanied by an absorption of energy. The energy discharged, at one time, is given by the formula hv , where h is the Planck constant and v is the frequency of the radiated energy.

quarter-wave plate. A compensator giving a retardation of about 130 nm, and a phase shift of 1/4 λ, thus constituting a device used with a polarizer and analyzer designed to produce circularly polarized light.

quartz. A transparent crystalline phase of SiO_2 with an index of refraction of 1.553 for the extraordinary ray (ε) and 1.544 for the ordinary ray (ω). It is a uniaxial mineral, positive in character, and belongs to the trigonal division of the hexagonal system of crystals. Crystalline quartz is sometimes used optically and microscopically and so is vitreous silica. See, *e.g.*, "quartz" slides and coverslips; "quartz" wedge.

"quartz" slides and covers. These are necessary for work with the ultraviolet microscope in the short wavelengths. However, for work with the 365 nm line, Corex® glass (*q.v.*) can be substituted and is much cheaper. "Quartz" slides and coverslips are vitreous silica.

quartz wedge. A compensator (*q.v.*) consisting of a gradual wedge of quartz (*q.v.*) of such orientation and dimensions as to show at least several orders of retardation (*q.v.*) colors as illustrated by the Michel-Lévy chart (*q.v.*) of retardation and birefringence (*q.v.*) *vs.* thickness of specimen. The Babinet (*q.v.*) compensator employs two opposing quartz wedges, calibrated in terms of retardation.

quick-change condenser. A condenser (*q.v.*) with a top lens mounted so that it is quickly removed to reduce its numerical aperture and widen the illuminated field. One method of quick change is to mount the top lens on a hinged arch.

radiation, monochromatic. Radiation at a single wavelength, and by extension, radiation of a very small range of wavelengths or frequencies.

Ramsden, Jesse (1735-1800). An English optician and inventor of the Ramsden ocular.

Ramsden circle (Ramsden disc, eyepoint, exit pupil). The circular spot of light formed at that distance above the eyepiece where the chief image forming rays cross, the back focal plane of the eyepiece. The objective back focal plane is in conjugate focus in this same plane. In visual microscopy, the point where the pupil of the eye is placed.

Ramsden eyepiece. An ocular consisting of two plano-convex lenses with the plane side of the lower lens nearer the objective. The focal plane is outside the system, hence it is of the positive type with the diaphram below the lenses. While this eyepiece gives a somewhat flatter field than the Huygenian and for this reason has been used in the past for micrometer oculars, the color correction is poor with the formation of color fringes around the object. The modern Ramsden eyepiece has an achromatic doublet for the eye lens to correct for color.

real image. See image, real.

reflection factor. The ratio of reflected light from a surface to the incident light. This is sometimes called the coefficient of reflection. Unless especially stated it takes into account both specular and diffuse reflection.

refraction. The change in direction of a ray of light in oblique passage from one transparent medium to another of different density, caused by the effect of a change of velocity of the light waves.

refractive index (n). The ratio of the velocity of light in a vacuum to the velocity in some medium. Refractive index generally increases with the atomic number of the constituent atoms.

refractive index by Becke line. With particles of the specimen mounted in a liquid of known refractive index relative refractive index is determined by (a) stopping down the condenser, (b) raising the focus of the microscope (_i.e._, from sharp focus move the objective away from the specimen), and (c) noting a halo or ring which will appear in the substance, liquid, or solid, with the higher refractive index. This halo is the Becke line, and the rule to remember is: "As the focus is raised, the Becke line moves toward and into the material of the higher refractive index."

relative aperture. The ratio of the focal length of a lens to the diameter of its entrance pupil. This gives a number known as the f-number, usually written f:8, f:16, etc. Thus, if the focal length is divided by the number 8, 16, etc., the result will be the diameter of the entrance pupil of the lens, or if the diaphram of the lens is wide open it will be very nearly the diameter of the free aperture of the lens. See f-number.

relative refraction, index of. The ratio of the refractive index of a substance to the index of its surrounding medium.

resin. A natural or man-made material that looks, feels, or behaves like rosin, shellac, or amber.

resinography. Morphology (*q.v.*) and structure (*q.v.*) correlated with composition or condition and with properties or behavior of resins (*q.v.*), polymers (*q.v.*), and their plastics.

resolution. The fineness of detail in an object which is revealed by an optical device. Objectively, resolution is specified as the minimum distance between two lines or points in the object that are perceived as separate by the human eye. Subjectively, the images of the two resolved points must fall on two receptors (rods or cones) which are separated by at least one other receptor on the retina (*q.v.*) of the eye. See resolving power and shape resolution.

resolving power. The ability to distinguish two small points in an image with their diffraction discs not overlapping more than half their diameter.

retardation. The actual distance of one of the doubly refracted rays behind the other as they emerge from an anisotropic substance. It depends on the difference in the two refractive indices, $n_2 - n_1$, and the thickness.

retardation plate. A plate placed in the path of a beam of polarized light for the purpose of introducing a difference in phase. Usually a quarter-wave plate and a first-order red plate are furnished with a polarizing microscope. See quartz wedge; compensator.

reticle (reticule, graticule). A group of lines, circles, angles, or some other pattern to be used as a measuring reference, as a focusing target, or to define a camera field of view.

reticle eyepiece. A microscope ocular having at its focal plane a reticle (*q.v.*).

retina. The sensory tissue of the inner side of the back of the eyeball. It is lined with tiny nerve endings called rods and cones. The image-forming rays from the eye lens are focused on the retina, and through the rods and cones, via the optic nerve, the sensation of vision is transmitted to the brain.

retinal image. The image formed by the lens of the eye on the retina (*q.v.*). The image is reversed and inverted.

retroflective screen. For viewing a projected image at a wide solid angle from the projecting beam, the viewing surface of the screen is made to scatter the image intact by means of solid, transparent beads, metallic paint, white pigment, or parallel ridges.

reversing prism. Used at the exit pupil to direct (project) the rays in a direction perpendicular to the microscope axis (as onto a viewing screen). The image is reversed.

rotating nosepiece. See nosepiece.

Rheinberg filter. A color-filter disk to be placed, as a darkfield stop would be placed, below the substage condenser. The central circular area, that is filled with one of the two or three color filters, should safely subtend the objective aperture. The annular quadrants around this are normally contrasting in color. This is/are the color(s) shown by the organisms or other specimen detail against the colored field. The effect is one kind of "optical staining" (*q.v.*).

rhombohedron. A crystal having 6 identical faces each of which is a rhombus.

rosin. A natural, unimeric (*q.v.*) resin (*q.v.*) from pine oleoresin. See colophony.

scanning electron microscope (SEM). An electron microscope in which the image is formed by a beam synchronized with an electron probe scanning the object. The intensity of the image-forming beam is proportional to the scattering or secondary emission of the specimen where the probe strikes it.

Schlieren system. A system for enhancing path length differences; refractive index differences in transparent media, or height differences for light reflected from surfaces.

screen. See optical light filter.

screw micrometer. See filar micrometer ocular.

SEM. Abbreviation for scanning electron microscope (*q.v.*).

semi-apochromatic objective. A compromise, in the correction for chromatic and spherical aberration, between achromatic and apochromatic objectives, such as a fluorite objective (*q.v.*).

Sénarmont compensator. See compensator.

shadow-cast replica. A replica which has been "shadowed". See shadowing.

shadowing. A process by which a metal or its compound is deposited on a specimen at an angle, from a heated filament in a vacuum, to enhance image contrast by inhibiting the deposition of the shadowing material behind projections.

sharpness. The visual impression of distinctness of detail in a photographic reproduction.

short-stop bath. An acid bath for photographic film or paper to stop developing action quickly. It is used as soon as the film is taken from the developer. Another function is to neutralize the alkalinity of developed film or paper before it goes into the acid fixer.

shutter. A device which permits regulation of the time during which light is allowed to act on a light-sensitive medium.

SI (Systeme Internationale). The International Systems of Units.

sign of double refraction. An empirical classification of crystals. It is positive for uniaxial crystals when $\varepsilon > \omega$, for biaxial crystals when $\gamma - \beta > \beta - \alpha$. It is negative for uniaxial crystals when $\varepsilon < \omega$, for biaxial crystals when $\gamma - \beta < \beta - \alpha$.

sign of elongation. Referring to the elongation of a substance in relation to refractive indices. If it is elongated in the direction of the high refractive index, it is said to have a positive sign of elongation. If it is elongated in the direction of the low refractive index, it has a negative sign of elongation; not to be confused with the sign of double refraction (*i.e.*, optic sign, *q.v.*).

Silverman illuminator. A small annular lamp used with a reflector, and mounted on the objective by means of a small three-jaw chuck. It furnishes above-stage illumination particularly suited to low powers. Because it generates considerable heat, care must be exercised in its use.

sine condition. The design of a lens must fulfill this condition, described by Abbe, if the image is to be aplanatic (*q.v.*). It states that the ratio of the sines of the angles of the incident and refracted rays to the axis must be constant; this constant is equal to the inverse of the magnification of the image.

skeletal. A single crystal grain, showing most rapid growth along body diagonals, like snowflakes, cf. dendritic.

slit microscope. See ultramicroscope.

slow ray. The slower of the two rays created by a crystal or fiber and the one that travels the path of higher refractive index.

Snell, Willebrord (1591-1626). Dutch mathematician and discoverer of the law of refraction which bears his name.

sodium lamp. A gaseous discharge lamp by the General Electric Company. Excellent as a source of essentially monochromatic yellow light but made up of the visible sodium doublet lines at 589.0 nm and at 589.6 nm on the wavelength scale.

specific optical dispersion. The difference between the refractive indices of light of two different wavelengths, both indices measured at the same temperature, the difference being divided by the specific gravity also measured in the same medium at the test temperature. For convenience, the specific dispersion value is multiplied by ten.

specimen. A piece or portion of a sample selected for examination. The specimen may, or may not be representative, whereas the sample may have been selected to be representative.

specimen chamber (electron optics). The compartment located in the column of the electron microscope in which the specimen is placed for observation.

specimen charge (electron optics). The electrical charge resulting from the impingement of electrons on a nonconducting specimen.

specimen contamination (electron optics). A change in the specimen caused by the condensation upon it of residual vapors in the microscope under the influence of electron bombardment.

specimen distortion (electron optics). A physical change in the specimen caused by desiccation or heating by the electron beam.

specimen grid. See specimen screen.

specimen holder (electron optics). A device which supports the specimen and specimen screen in the correct position in the specimen chamber of the microscope.

specimen screen (electron optics). A disk of fine screen, usually 200-mesh stainless steel, copper, or nickel, which supports the replica or specimen support film for observation in the microscope.

specimen stage (electron optics). The part of the microscope which supports the specimen holder and specimen in the microscope, and can be moved in a plane perpendicular to the optical axis from outside the column.

spectrophotometry. The quantitative measurement of reflection or transmission properties as a function of wavelength.

spectrum, secondary. The color or colors left uncorrected in a combination of lenses.

specular. Having a smooth mirror-like surface; by extension, the action of such a surface on light, as a specular reflection, meaning mirror-like reflection as opposed to diffuse reflection. Also used in light transmission, as specular transmission, denoting transmission of a pencil of light without diffusion.

sphenoid. A crystal form made up of two nonparallel faces not symmetrical about a plane of symmetry.

spherical aberration. See aberration, spherical.

spherulite. A sphere composed of needles or rods all oriented perpendicular to the outer surface, or a plane section through such a sphere.

stage. A device for holding a sample in the desired position in the optical path.

stage, mechanical. A small fixture, either built into the light microscope stage or attached separately; it holds the specimen slide and has two horizontal screw adjustments at right angles to each other. The screw motions permit the specimen to be moved as desired. The quantitative type has vernier scales for reading the amount of displacement to 0.1 mm. This stage is sometimes called the traversing stage.

stage micrometer. A graduated scale used as a standard on the stage of a light microscope for calibrating an eyepiece micrometer; also for determining the magnification of a set-up in photomicrography, etc.

staining, optical. See optical staining.

star test. This test is well described in Shillaber's text[10]. It affords a means for determining whether under- or over-correction exists in the microscope system. Its intelligent use helps ensure attainment of the optimum image of which an objective is capable.

stereomicroscope. A microscope (simple or compound) for each eye (binocular), giving different aspects and, therefore, a stereoscopic effect. There are two kinds of compound stereomicroscopes: binobjective and common main objective (CMO). See Greenough.

stop. See diaphram.

stop, darkfield. See central stop.

structural color. Iridescence or interference colors due to a thinly layered surface, transparent interface, or periodic structure.

structure. The mode of construction of an animate or inanimate body or system from units such as atoms, ions, molecules, cells, crystals in a fluid, plastic, or solid state. *Cf.* morphology.

submicroscopic. Particles which, although visible in the ultramicroscope (*q.v.*), are too small to be resolved by visible light. This places their size between 0.2 µm and 0.005 µm.

tablet. A shape having opposite faces at least 4X but less than 10X greater than the other faces.

temperature coefficient of refractive index. The change in refractive index (n) with temperature. The degree of variation of n depends on the composition of the substance and the state of aggregation, *e.g.*, whether it is a solid or a liquid. It is usually about 100X larger for liquids than for solids; about -0.0005/°C for liquids.

tomography. X-ray photography by which a single plane is photographed. A method of collecting data from various angles. Combined with a computer it becomes CAT (*q.v.*): Computer-aided tomography. See also microtomography.

trademark. Abbreviated TM or, if registered in the U.S. Patent Office, signified by [R] or ®. A trademark is a mark, symbol, or word (non-descriptive) that is capitalized, *e.g.*, Fiberglas® for fibrous glass by Corning Glass Corp.; Polaroid® registered by Polaroid Corporation for its cameras, photographic material, or polarized film.

trade name. (Two words, not capitalized) for a name used in the trade by anyone (*e.g.*, chrome for chromium plate; bakelite for phenol-formaldehyde plastic; rayon for regenerated cellulose fiber). Not to be confused with trademark (*q.v.*).

translucent. Transmitting of light in such a way that image-forming rays are irregularly refracted and reflected.

transmission microscope. A microscope in which the image-forming rays pass through (are transmitted by) the specimen being observed. Refers to both light and electron microscopes.

transmitted light. The usual method for illuminating transparent microscopic specimens. The light is concentrated on the specimen by the substage condenser. Objects appear in outline (refraction images) or colored on a bright field (color images).

transparent. Any optical substance or other material which is so optically clear that a pencil of light passing through it may be focused without undue effect on the image. A piece of clear glass is transparent; a piece of opal glass is translucent.

triplet. A combination of three simple lenses cemented together. Two positive lenses with a negative lens between them can produce a well-corrected system.

tubelength adjustment. The drawtubes of some light microscopes can be extended beyond or pushed within the standard setting demanded by the objective in use. It is normally set at 160 mm. Extending the tube beyond the normal setting produces an over-correction in the microscope system and is useful to offset under-correction due to various causes. Conversely, shortening the tubelength tends to correct for an over-corrected system. Modern microscopes generally do not have drawtubes.

tubelength, mechanical. This distance is measured from where the objective screws on to where the eyepiece fits in. The American standard mechanical tubelength is 160 mm. For Leitz objectives it was once 170 mm.

tubelength, optical. This distance is measured from the upper focal plane of the objective to the image formed by the objective alone upon removal of the eyepiece.

Tyndall blue. The blue color produced by light scattered by extremely fine particles suspended in a liquid or a gas. Often observable through the darkfield microscope. Also called the Tyndall effect.

twin-plane. The common-composition plane of two contacting crystals. It must have the same crystallographic indices for each and every crystallographic instance. (Fortuitous sharing of various types of crystallographic planes is not "twinning".)

ultramicroscope. A light microscope so arranged that the specimen, usually a solid-liquid colloid or suspensoid, is illuminated by a strong pencil of light at right angles to the microscope axis. The visibility is limited by the intensity of the light source. A laser can be used as the source. It is used to detect the presence of light-ultramicroscopic particles within the range of about 0.005 μm to 0.2 μm. Not to be confused with an electron microscope.

ultramicroscopic. A term applied to particles less than 0.1 μm in diameter, hence too small to be truly resolved by the light microscope. Under the ultramicroscope (*q.v.*) they look like stars in the sky. Their differences in size are merely indicated by differences in brightness.

ultraviolet radiation. That part of the electromagnetic spectrum extending from 12 to 400 nm. For photomicrography the most useful wavelengths are 365 nm and 275 nm (nanometers). The eyes should be protected from ultraviolet radiation at all times!

ultraviolet microscope. A microscope employing invisible ultraviolet "light", throughout: condenser, object slide and coverslip, objective, and "ocular". The ultraviolet image is invisible but is recorded on photosensitive film.

undulose extinction. Nonuniform extinction of a substance between crossed polars. The areas of complete extinction move progressively with a fanlike motion across the surface of the substance as the stage is rotated.

uniaxial crystals. Anisotropic crystals in the tetragonal and hexagonal systems having one unique crystallographic direction and either two (tetragonal) or three (hexagonal) directions which are alike and perpendicular to the unique direction.

unimeric. A single, large molecule that is not monomeric, oligomeric, or polymeric. Rosin (_q.v._), for example, consists primarily of unimeric, tricyclic, monocarboxylic acids of the general empirical formula $C_{20}H_{30}O_2$.

unpolarized light. A bundle of light rays having a common propagation direction but different vibration directions.

vertical illumination. Brightfield illumination by light from the objective which is reflected or scattered from the (usually opaque) object. Illumination is by means of a vertical illuminator placed above the objective. Light is brought into a side tube and directed toward the back aperture of the objective by a tiny mirror or prism, or else by a full-aperture transparent-reflector (thin glass plate) 45° to the axis of the bodytube.

virtual image. Such as seen in a mirror or through a magnifier. A virtual image has no real existence in space as does a real image from a lens. It does have a definite location, however, caused by the angles of divergence of the rays received by the eye. This can be shown by the common school experiment of placing a pin coincident with its mirror image behind a sheet of glass acting as a partial mirror. Its location can also be placed in design by extrapolating backwards to a focus. If a magnifier is used as it should be, with the object at its focus, the virtual image is at infinity. The same is true for a microscope focused for the relaxed eye. See distance of virtual image.

visibility, limit of. For the normal eye, the limit of visibility is considerably below the limits of resolution (_q.v._). It depends largely on contrast and intensity of illumination.

visual angle. The angle at the eye subtended by the limits of the object. If the object is at the threshold of resolution, the angle is said to be 1 minute of arc.

vitreous humor. An aqueous jelly-like mass occupying the space between the eye lens and the retina.

wavefront. A surface at which all vibratory motion is of like phase concurrently.

wavelength. In wave motion, the distance between a point on one wave and a corresponding point on the next wave, *e.g.*, the distance between the crest of one wave and the crest of the next wave. The nominal wavelength of red light is taken as 700 nm; for violet light, 400 nm.

wave number. The number of waves or cycles of light flux or radiant energy, measured through a distance of 1 cm.

widefield eyepiece. An ocular with an achromatic doublet for the eyelens and with the plane side of the lower lens nearest the objective. Such a corrected system does not have to be stopped down with a diaphram, hence a large flat field is achieved.

working distance, free. The distance between the front lens of the objective and the coverslip (or uncovered object) when the lens is focused on the specimen.

x-x axis. The horizontal axis, or axis in the left to right direction, in a plane Cartesian coordinate system along which a row of functional patterns is nominally disposed by stepping and repeating.

xylene. One (or all) of the three isomers of $C_6H_4(CH_3)_2$, ortho, meta, para. The commercial mixture, often called xylol, is generally used in microscopy for cleaning lenses, etc.

y-y axis. The vertical axis orthogonal to the x-x axis in a plane Cartesian coordinate system along which a column of functional patterns is nominally disposed by stepping and repeating.

BIBLIOGRAPHY

1.   *Glossary of Microscopical Optical Terms and Definitions*, New York Microscopical Society, 15 West 77th Street, New York, N.Y. 10024, 73 pages, First Edition, (1978).

2.   *Compilation of ASTM Standard Definitions*, 1986 edition, American Society for Testing and Materials, 1916 Race Street, Philadelphia, PA 19103.

3.   ASTM designation E7-83, *Standard Definitions of Terms Relating to Metallography*, 1983, American Society for Testing and Materials, 1916 Race Street, Philadelphia, PA 19103.

4.   ASTM designation E375-85 (Reapproved 1980), *Definitions of Terms Relating to Resinography*, 1980, American Society for Testing and Materials, 1916 Race Street, Philadelphia, PA 19103.

5.   ASTM designation E175-82, *Standard Definitions of Terms Relating to Microscopy*, 1982, American Society for Testing and Materials, 1916 Race Street, Philadelphia, PA 19103.

6.   The Optical Data Processing Family Tree, article by Joseph W. Goodman, in *Optics News*, May/June 1984, p. 25, ff.

7.   *Glossary*, 1985, *Polarized Light Microscopy*, McCrone Research Institute, 2820 South Michigan Avenue, Chicago, IL 60616.

8.   ASTM designation E20, *Particle Size Measurement, Microscopical Methods, Recommended Practice*, American Society for Testing and Materials, 1916 Race Street, Philadelphia, PA 19103.

9.   ASTM designation E-211-82, *Standard Specification for Cover Glasses and Glass Slides for Use in Microscopy*, American Society for Testing and Materials, 1916 Race Street, Philadelphia, PA 19103.

10.   C. D. Shillaber, *Photomicrography*, 1944, John Wiley & Sons, Inc., New York, NY 10046.

11.   R. P. Loveland, *Photomicrography*, Vols. 1 & 2, first printing, 1970, John Wiley & Sons, Inc., New York, NY 10016, second printing, 1985, R. E. Krieger, Malibar, FL 32950.

12.   T. G. Rochow and E. G. Rochow, *An Introduction to Microscopy by Means of Light, Electrons, X-Rays, or Ultrasound*, 1978, Plenum Press, New York, NY 10013.

13.   T. G. Rochow, *Light Microscopical Resinography*, 1983, Microscope Publications, Chicago, IL 60616.

14.   T. G. Rochow and E. G. Rochow, *Resinography*, 1976, Plenum Press, New York, NY 10013, including its Glossary, pp. 165-184.

APPLICATION FOR MEMBERSHIP

Application for membership should be sent to:

New York Microscopical Society
Chairman, Membership Committee
15 West 77th Street
New York, New York 10024

Date of First Reading _______________ .
Date of Second Reading _______________.

I hereby apply for membership in the NEW YORK MICRO-
SCOPICAL SOCIETY in the class which I have checked below:

_________ Junior Member (under 21) - - - - - - - - - $ 5.00
_________ Annual Member - - - - - - - - - - - - - - - $ 20.00
_________ Supporting Member - - - - - - - - - - - - $ 40.00
_________ Corporate Member - - - - - - - - - - - - $100.00
_________ Life Member (payable within the year) $200.00

I agree to further the interest of the Society to the best of my
ability.

Name____

Home Address ____

City, State, Zip ____

Business Address ____

City, State, Zip ____

Home Tel. No. Area Code (_______) ___________________________

Bus. Tel. No. Area Code (_______) ___________________________

MY DUES FOR THE FIRST YEAR ARE ENCLOSED

Proposed By _________________ Approved_____________________

Elected _______________ Date _______________

Please list personal data for our files.

Date of Birth _______________ (If over 21, "Adult" may be written)

Academic and Honorary Degrees, if applicable:

Degree   Conferring Institution   Date

___________   ___________________   __________

___________   ___________________   __________

Business or Institutional Connection and in what Capacity

____

Membership in Scientific Societies_______________________________

____

Principal work or interest in microscopy

____

Contributions (if any) to Scientific Literature:

____

____

On what topic, if any, are you available as a speaker?

____

What types of programs would you like to have presented?

____

Are you interested in working on committees, or in other active
participation?
_________ Yes   _________ No